"Han Lin's book on China's efforts to deal with carbon emissions is remarkable for her insider's view of China's cultural and historical context. Exhaustively researched through interviews with significant actors, online forums and policy documents, the book shows why China changed its stance on climate change and provides recommendations on how China can optimize its energy policies to achieve better outcomes in climate change mitigation. Among the book's many contributions are Dr. Lin's attention to the role of Confucian tradition in shaping the importance of hierarchy, her attention to the role of the expert, and the importance of government-linked think tanks in the policy process. Bureaucratic rivalries and China's recent reallocations of responsibility are well explained, as is China's choice of emissions trading over carbon tax. The book is indispensable for anyone seeking a deep understanding of China's climate change policies".

– Judith Shapiro,
American University, author of China's Environmental Challenges

Energy Policies and Climate Change in China

In the face of growing environmental challenges, including climate change and energy security, countries across the globe are developing new policies and programs to address these challenges, and China is no exception. This book analyses China's two most significant climate-related energy policies, the Clean Development Mechanism (CDM; including the later Chinese Certified Emission Reduciton – CCER) and the Energy Conservation and Emission Reduction Scheme (ECERS).

This work specifically examines the strengths and weaknesses of these policies to highlight the deficiencies and advise how they can be optimised, so China can better achieve its emission reduction goals. It analyses the roles and relationships between relevant actors and identifies how successful their cooperation has been, and what factors have affected it. Importantly, the work draws on a wide range of sources from central ministries to civil society, including interviews with Chinese officials, scholars, energy company managers, environmental non-governmental organisation (ENGO) personnel, media reports, and online forum discussions. In doing so, the book not only analyses the thoughts of policymakers, as many works do, but also those implementing the policies and those impacted by the policies. The book concludes by offering detailed and practical solutions to address each specific deficiency in the CDM and ECERS policies, with the aim of providing innovations and alternative approaches to improve current and future policies in China.

This book will be of great interest to students, scholars, and policymakers interested in climate change, energy, and Chinese environmental policy and politics.

Han Lin holds a PhD from the College of Business, Government and Law, Flinders University, Australia, and currently teaches at Flinders University.

Routledge Studies in Energy Policy

Making Electricity Resilient
Risk and security in a liberalized infrastructure
Antti Silvast

Energy Policy in China
Chi-Jen Yang

Fossil Fuel Subsidy Reforms
A Guide to Economic and Political Complexity
Jun Rentschler

Low Carbon Politics
A cultural approach focusing on low carbon electricity
David Toke

Governing Shale Gas
Development, Citizen Participation and Decision Making in the US, Canada, Australia and Europe
John Whitton, Matthew Cotton, Ioan M. Charnley-Parry and Kathy Brasier

Business Battles in the US Energy Sector
Lessons for a Clean Energy Transition
Christian Downie

Guanxi and Local Green Development in China
The Role of Entrepreneurs and Local Leaders
Chunhong Sheng

Energy Policies and Climate Change in China
Actors, Implementation, and Future Prospects
Han Lin

For further details please visit the series page on the Routledge website: http://www.routledge.com/books/series/RSIEP/

Energy Policies and Climate Change in China

Actors, Implementation, and Future Prospects

Han Lin

Routledge
Taylor & Francis Group
LONDON AND NEW YORK

earthscan
from Routledge

First published 2020
by Routledge
2 Park Square, Milton Park, Abingdon, Oxon OX14 4RN

and by Routledge
52 Vanderbilt Avenue, New York, NY 10017

Routledge is an imprint of the Taylor & Francis Group, an informa business

British Library Cataloguing-in-Publication Data
A catalogue record for this book is available from the British Library

Library of Congress Cataloging-in-Publication Data
A catalog record has been requested for this book

ISBN: 978-0-367-00173-5 (hbk)
ISBN: 978-0-429-40097-1 (ebk)

Typeset in Bembo
by Cenveo® Publisher Services

MIX
Paper from responsible sources
FSC® C013056

Printed and bound in Great Britain by
TJ International Ltd, Padstow, Cornwall

Contents

Acknowledgements

This book started life as my PhD thesis. I would like to thank my supervisors Associate Professor Cassandra Star and Dr Zhang Zhibin for their help and support during my PhD candidature.

My editors at Routledge Hannah Ferguson and Matthew Shobbrook are extremely professional and helpful. They responded quickly to my questions and made the whole publication process run smoothly.

I sincerely thank my interviewees from National Development and Reform Commission of the PRC, Ministry of Environmental Protection (Now Ministry of Ecology and Environment) of the PRC, Ministry of Transport of the PRC, the Chinese Academy of Social Sciences, and a range of environmental non-governmental organisations and industries. All of them provided me with invaluable information for my research. Without their willingness to participate, this project would not have been as comprehensive.

Finally, I would like to thank my family for their unconditional support. My husband Jeff took a personal interest in my book project. He proofread the book and gave lots of valuable comments. My parents helped look after my children while I was writing my book. My children Alexander and Nicholas brought me a lot of joy and welcome breaks during the writing process. Without the support my family it would be a "mission impossible" to complete my book.

Fate brought me two precious little angels during the construction of this book. My first son was born in the first year of my candidature, and my second son was born in the second last year of my candidature. So, it was a candidature full of challenges and excitement.

1 Climate change, energy security, and China's stance

China's stance in international climate change negotiations

Facing the threats of global climate change, a range of international mitigation attempts were carried out by multilateral cooperation over several decades. As early as the 1980s, the scientific concern of global warming was developed, and this set out the scientific foundation for future climate change mitigation efforts. From 1985 to 1988, climate change was transformed from a scientific issue into a policy issue through a range of international conferences and the agenda of climate change mitigation was established. Since then, governments started to get involved in the negotiation process for climate change mitigation (Bodansky 2001, pp. 23-24). China started to engage more in international environmental affairs during this period as its reform and opening up deepened (Chen 2012, p. 1). China's participation in and attitude towards climate change would have a significant impact on the outcome of climate stability due to its large population and rapidly developing economy (Schreurs 2011, p. 449). In 1992, China signed and ratified a non-binding agreement at the Rio Earth Summit which aimed at stabilising greenhouse gas emissions at 1990 levels through individual country's specific policies and measures (Mckibbin & Wilcoxen 2002, p. 41). The Rio Earth Summit was seen as a landmark for the earliest international negotiation on the global climate change policy (Bodansky 2001, p. 23).

However, the fact that China ratified the Rio Earth Summit did not mean that China took a proactive approach towards climate change mitigation. Instead, China proposed the principle of "common but differentiated responsibilities" towards climate change mitigation (Kobayashi 2003, p. 88). It argued that developed countries should bear responsibility for climate change and should aid developing countries in financial and technological terms to facilitate mitigation and adaptation (Chen 2012, p. 5; Wu 2013, p. 780). Such an argument had earned broad support from developing countries.

It was, however, very unfortunate that few significant policies had been successfully implemented after the Rio Earth Summit and greenhouse gases emissions continued to soar (Höhne et al. 2003). In an attempt to address the deficiencies of the United Nations Framework Convention on Climate Change (UNFCCC) at the Rio Earth Summit, a range of "conference of the parties" (COP) meetings were determined for further international cooperation on climate change mitigation.

In December 1997, the COP 3 was held in Kyoto and the Kyoto Protocol was drafted. The Kyoto Protocol was the first international treaty that set detailed legally binding obligations for industrialised countries to cut greenhouse gases emissions (Chen 2012, p. 7). The protocol set up emissions targets for 39 developed countries which aimed for 5% emission reduction from 1990 levels within the next 14 years. The protocol included three mechanisms: international emissions trading (IET), joint implementation (JI), and the Clean Development Mechanism (CDM) (UNFCCCa n.d., n.p.). China, as a developing country, was not required to meet specific greenhouse gas reduction targets under the Kyoto Protocol like the industrialised countries, but it was welcomed to participate in the emission offsetting CDM (Chen 2012, p. 7).

Although China was not required to meet any legally binding emission reduction target under the Kyoto Protocol, it did complete its domestic ratification in 2002 by setting up national emissions inventories for reporting national programs and exchanging information with other nations. The publication of the "National Climate Change Program" confirmed the seriousness of global climate change, which was an initial sign that the Chinese government started to treat climate change seriously at the national level (NDRC 2004, p. 7). Moreover, China actively participated in regional climate change mechanisms such as the Asia-Pacific Partnership on Clean Development and Climate, which focused on voluntary participation and energy efficiency improvement. These actions set out the contrast between China and the United States, which had withdrawn from the Kyoto Protocol (Heggelund & Buan 2009, p. 302)

The Kyoto Protocol was a turning point where China's attitude towards global climate change mitigation switched from passive to proactive. There are multiple reasons behind this attitude switch, ranging from the domestic energy challenge (Yao & Chang 2014), to the public's voice in treating air pollution (Ho & Nielsen 2007), to the concern over establishing a positive international image (Yu 2008; Zhang 2004). Among all these reasons, the domestic energy shortage and concern for energy security was the most powerful driver, as argued below.

Behind the glamour of China's rapid economic development lies the shadow of its energy crisis. China constantly experienced energy shortages since its reform in the late 1970s, except for a couple of years after the 1997 Asian Financial Crisis when energy supply could basically meet demand (Qi 2013, p. 128). In 2002, China's energy shortage intensified after recovering from the financial crisis. In 2004, 24 provinces out of 31 experienced black-outs, causing enormous disturbance and loss to the economy (Lin 2007, p. 916). Former premier Wen Jiabao identified the shortage of coal, electricity, petroleum, and transportation as being some of the most pressing problems at the *Third Session of the 10th National People's Congress* on 5 March 2005, where the most important and urgent domestic issues were proposed and discussed (Wen 2005, n.p.).

China has been heavily reliant on imported oil for a few decades, which is a potential threat to the nation's energy security (Zhang 2011, p. 7612). The existing high-energy input style industries, inherited from the Soviet Union, further worsen the problem (Mackerras, Taneja & Young 1998, p. 149). As a result, a demand for efficiency improvement emerged in China. Although China has relatively rich coal reserves and currently coal serves approximately 70% of China's total energy needs (Wu & Zhang 2016, p. 126), this does not provide complete relief for its energy outlook. This is because coal mines are dispersed unevenly across the country, being mainly located to the west and the north, while the major high-energy demand regions are in the south and east along the coast. This has caused coal-fired power plants to fail to deliver electricity from time to time during energy demand spikes. The 2008 snowstorm in southern China was such an example. Railways were out of use due to the weather conditions, trucks loaded with coal were stuck on the frozen highways and, as a result, the urgently needed coal was not able to be delivered to power plants to generate electricity for essential heating (Zhou 2010, p. 3756).

Apart from the concerns about domestic energy shortages, China also cared much about its international image. Acquiring a positive image through involvement in international climate change negotiation is another driver behind China's attitude switch. In the report *China's Policies and Actions on Climate Change 2014*, it was clearly stated that:

> Pursuing green, low-carbon development and actively addressing climate change is not only necessary to advance our [China's] ecological progress and put our development on a sustainable path, but will also demonstrate to the world that China is a responsible country committed to making an active contribution to protecting the global environment. (NDRC 2014, p. 1)

Zhang (2004, p. 78) also pointed out that China used climate change to both strengthen its relationship with other developing countries during early negotiations by proposing "common but differentiated responsibilities" and to enhance its relationship with developed counties in the later stage by actively participating in the CDM. This can be reflected from the website of the National Development and Reform Commission (NDRC), the central ministry responsible for climate change and energy policy development. On the Chinese version of the site, the home page is dominated by a range of important national economic plans, strategies, and projects. While on the English version of the site, the home page contains six news releases from 2013 to 2015, with five of them related to climate change (NDRC 2016, n.p.).

The dramatic difference in contents between the two versions of the websites further demonstrates that a positive international image is what the Chinese government actively pursues, and global climate change mitigation provides an important opportunity for this.

Domestically, the voice of the Chinese people demanding clean air is another reason for the Chinese government to reconsider its stance on climate change. This is because proactive climate policies can bring significant co-benefits towards the treatment of air pollution (Aunan, Mestl & Seip 2003, p. 287). In early 2015, Chai Jing, a former China Central Television (CCTV) host, presented a 103-minute self-funded documentary *Under the Dome*. It discussed the causes of city smog, its impacts on public health, and provided several suggestions to deal with it. This documentary attracted huge attention from the public, and started a national conversation on the issue of air pollution. The documentary received 31 million clicks within 24 hours on six mainstream video sites including qq, youku, sohu tv, letv, tudou, and qiy (360doc 2015, n.p.). Even the minister-in-charge of the Ministry of Environmental Protection (MEP), Chen Jining, sent Chai Jing a thank-you message after watching the documentary. Although there were mixed comments including both praise and complaints, it was undeniable that Chai Jing's documentary was a summation of the Chinese public's increasing environmental awareness. It was also very significant that the public used such media as a tool to express their concerns to the government in a national manner.

Due to the influence of these factors, China continued its commitment to reduce emissions on a voluntary basis, but the bottom line was that China did not accept the imposition of an absolute cap on the quantity of emission. Even during the later negotiation stage when China committed to more ambitious targets in combating climate change, it still firmly held this bottom line. For instance, at COP 15, Copenhagen, 2009, China announced the goal of cutting its carbon intensity by 40–45% below the 2005 level by 2020 (BBC 2009, n.p.). Again, at the COP 21 Paris Summit, China committed to a 60–65% carbon intensity reduction from the 2005 level by 2030 (NDRC 2016, n.p.). Carbon intensity reduction is related to how efficiently fossil-based energy is used rather than the total quantity that is consumed, and therefore still allows emission growth even if greater efficiency is achieved.

China's stand is understandable and deserves sympathy in the sense that it faces not only a climate challenge but also an energy challenge. With the world's largest population and a coal-dominated energy structure, China faces possibly the greatest difficulty in effectively mitigating climate change and simultaneously guaranteeing sufficient energy supply for its growing economy. Population growth, on-going industrialisation, a manufacturing and exporting dominated economic pattern and people's demand for improved standards of living have all driven up the energy demand (Guan et al. 2008, 2009; Wei et al. 2007; Zweig & Bi 2005).

China became an oil importer in 1993 (Lheem 2008, p. 173). It accounts for 41% of the world's oil demand growth and is currently the largest net oil importer in the world (Tian 2016, p. 36). China's acquisition of overseas oil in the Middle East, North Africa, Russia, and Central Asia can temporarily relieve its domestic energy shortage, but it cannot fully guarantee the nation's future oil supply. This is because, first, the reality must be faced that the world is fast approaching the peak of many fossil fuels. Some argue that peak oil is very close (Aleklett et al. 2010; Leder & Shapiro 2008). Some also argue that since 2015 the world's conventional crude oil production can no longer match the demand, and thus crude oil has already peaked (Murray & King 2012, p. 433; Pennock, Poland & Hancock 2016, p. 180). Nonetheless, this indicates that the world's oil reserves are fast depleting. Peak gas is also not far away, possibly arriving in just a few decades' time (Laherrère 2004), and peak coal is predicted to arrive before 2050 (Mohr & Evans 2009). Second, China's acquisition of overseas energy resources also imposes potential diplomatic or even military conflicts between China and the United States of America. This is because many of the countries China buys oil from are countries that the United States of America tries to isolate, such as Iran, Sudan, and Burma (Lheem 2008, p. 182). Such countries are usually viewed as potential threats within the USA's sphere of influence. The United States of America fears that China's involvement in such regions will challenge its military dominance and break the balance of power (Zweig & Bi 2005, p. 26).

Securing sufficient energy resources has been seen by many as one of the most important conditions for national security (Raphael & Strokes 2016; Stringer 2008; Wu & Storey 2007; Yergin 2006). Viewing energy security from a traditional geopolitical standpoint, serious competitions or even conflicts among nations over scarce energy resources are unavoidable due to the limited energy stocks and the world's increasing energy needs (Garrison 2009, p. 1; Griffin 2009, p. 11; Goldthau & Witte 2010, pp. 1-3; Nyman 2015). This argument is accepted and embraced by the Chinese government to a significant extent in its traditional energy policy development, as it reflects the reality that China is in a zero-sum situation against other nations when it comes to searching for overseas energy resources. Zweig and Bi (2005, pp. 26-27) described the situation this way:

> Although China's new energy demands need not be a source of serious conflict with the West in the long term, at the moment, Beijing and Washington feel especially uneasy about the situation. While China struggles to manage its growing pains, the United States, as the world's hegemony, must somehow make room for the rising giant; otherwise, war will become a serious possibility.

Realising the limits and possible conflicts caused by the acquisition of overseas energy resources, the Chinese government adopted a range of other measures for energy security, such as those dealing with energy efficiency improvement and development of renewable energies (Glomsrød & Wei 2016; Mathews & Tan 2014).

Energy policies within the context of climate change

This alternative path towards energy security is the focal point of this book. It explores China's energy policy options within the context of climate change by asking how China can optimise its current energy policies in order to achieve better outcomes in climate change mitigation. To do so, this book will evaluate China's climate-related energy actors and two important climate-related energy policies, namely the CDM (including Chinese Certified Emission Reduction Program, CCER) and the domestic Energy Conservation and Emission Reduction Scheme (ECERS). The reason why CDM and ECERS are chosen is that they are the two largest scale and most widespread climate-related policies ever carried out in China, with one residing in the international context focusing on the market approach, and the other in the domestic context, mainly using the administrative approach. The timeframe of the two policies roughly coincides with each other (2006–2020); therefore, the two policies can demonstrate China's major energy and climate achievements and drawbacks within the same era.

CDM, as a component of the Kyoto Protocol, allowed Annex I developed countries[1] to invest in greenhouse gases emission reduction projects in developing countries, so that developing countries can benefit from sustainable development, and investors can obtain Certified Emission Reductions (CERs) to offset their emissions or to trade on the international market (The World Bank 2004, n.p.). Currently, there are over 8000 CDM registered projects worldwide, representing approximately US$300 million investment in clean energy and emission reduction projects. In 2014, CDM projects achieved 300 million tons of CO_2 equivalent emission reduction, which accounted for 1% of the total global emissions (Hone 2017, n.p.). Hence, without CDM, the world's total emissions would be 1% higher. Unfortunately, the future of CDM is uncertain under the Paris Agreement (2016) due to a significant drop in the CER price. CDM might no longer hold its original status as an important international mitigation mechanism. However, the significance of CDM to China is longer lasting than the life-span of the mechanism itself. This is because the Chinese government regards the true value of CDM as introducing the concept of a carbon market and market mechanism in climate change mitigation. The fluctuation of the CER price had only a minor impact on the government's belief in the benefits of CDM. By 23 August 2016, the Designated National Authorities (DNA) of China had approved 5074 CDM projects. China is so far the world's largest producer of CERs, generating over 50% the world's total (Department of Climate Change, NDRC 2016, n.p.). China's domestic carbon offsetting program, CCER, developed based on CDM, is now implemented in China's domestic ETS as a successful component. In this sense, CDM not only introduced the concept of mitigating climate change through a market mechanism but also assisted China's sustainable development, and contributed greatly to energy efficiency improvement and renewable energy development.

Among all of China's domestic climate-related energy policies, ECERS is of the greatest significance. ECERS is a nationwide scheme that aims to achieve energy conservation and emission reductions in all sectors, including the energy-generation sector, the manufacturing sector, the building sector, the transportation sector, and the agriculture and forestry sector (The Central People's Government of PRC 2006a, n.p.). It was initially proposed in China's 11th Five-Year Plan for National Economy and Social Development. It is a long-term scheme which had been carried out through the 11th Five-Year Plan (2006–2010), the 12th Five-Year Plan (2011–2015) and is now being implemented for the 13th Five-Year Plan (2016–2020). ECERS, as a domestic climate-related energy policy, coincides with the international CDM in terms of time frame and the focus on energy efficiency improvement and renewable energy.

ECERS achieved overall positive results. From 2006 to 2015, China cumulatively forced the closure of approximately 100 million kW small thermal power generation capacity (NEA 2015, n.p.). A range of advanced energy-saving technologies were applied to qualified conventional coal-fired power plants, and the net coal consumption per each kWh of electricity generation dropped by 38 gm, which was approximately a 10% reduction (Qi 2013, p. 143; Wang 2015, p. 87). This achievement had a significant meaning to China's carbon intensity reduction and overall carbon emission reduction due to the large proportion of coal power in China's energy mix. During the 11th Five-Year Plan (2006–2010), the national energy intensity per unit of Gross Domestic Product (GDP) dropped 19.06% (NEA 2011, n.p.). During the 12th Five-Year Plan period (2011–2015), again ECERS achieved over 20% reduction in energy intensity which exceeded the planned 16% (Xu 2016, n.p.). The 13th Five-Year Plan (2016–2020) aims to achieve another 15% in energy intensity reduction on the 2015 level.

The development of renewable energies was also rapid under ECERS. During the 11th Five-Year Plan, China's wind power capacity increased from 1.27 GW in 2005 to 44.73 GW in 2010, which increased over 35-fold (Kang et al. 2012, p. 1907). During the 12th Five-Year plan, wind power capacity increased fourfold (NEA 2016, n.p.). During the same period, China's solar energy capacity increased 168-fold (NEA 2016, n.p.). By 2014, China surpassed Japan and the United States to become the world's largest photovoltaic (PV) modules consumer (iiEnergy 2016, n.p.). In fact, the proportion of China's renewable energy already reached 17% of the national energy mix by mid-2015, which was earlier and faster than the planned 15% of renewable energy in national energy structure by 2020 (Xinhua News 2015, n.p.).

CDM and ECERS are of great significance as they were designed to address China's energy security and climate change issues simultaneously. However, are they the most effective policies? Are there any weaknesses associated with them? If there are, how can the weaknesses be addressed? In order to answer these questions, this book will carry out analyses on two levels. The first-level analysis will focus on relevant actors and their strengths, relationship, and inputs in the development of China's climate-related energy policies. The second-level analysis will focus more on the implementation of the two important policies mentioned above. The investigation of their weaknesses is significant for understanding their overall effectiveness. Following the analyses, it will follow up on policy updates during the 13th Five-Year Plan to see whether the identified policy weaknesses are addressed, and if not, it will provide implications for addressing these.

Plan of the book

This book consists of seven chapters including this introductory chapter. Chapter 2 develops the conceptual framework by exploring the key concepts of policy approaches relevant to China. Chapters 3, 4, and 5 are the major analysis chapters. Chapter 3 deals with the actors in China's climate-related energy policies and their impact on policy development. Chapter 4 analyses the strengths and weaknesses of CDM in China, and Chapter 5 investigates the strengths and detrimental side effects associated with ECERS. After this major analysis, Chapter 6 investigates policy updates and puts forward implications to address the weaknesses raised in Chapters 3, 4, and 5. Chapter 7 concludes the book.

Note

1. List of Annex I countries: Australia, Austria, Belarus, Belgium, Bulgaria, Canada, Croatia, Cyprus, Czech Republic, Denmark, Estonia, European Union, Finland, France, Germany, Greece, Hungary, Iceland, Ireland, Italy, Japan, Latvia, Liechtenstein, Lithuania, Luxembourg, Malta, Monaco, Netherlands, New Zealand, Norway, Poland, Portugal, Romania, Russian Federation, Slovakia, Slovenia, Spain, Sweden, Switzerland, Turkey, Ukraine, United Kingdom of Great Britain and Northern Ireland, United States of America (UNFCCC) <http://unfccc.int/parties_and_observers/parties/annex_i/items/2774.php>.

References

360doc 2015, *Shishang zuiquan Chaijing <cangqiong zhixia> shijian shimo yu gefang shengyin (The most comprehensive information about Chaijing's <Under the Dome> and comments from all sectors)*, viewed 19 May 2017, <http://www.360doc.com/content/15/0302/13/21856245_451946747.shtml>.

Aleklett, K, Höök, M, Jakobsson, K, Lardelli, M, Snowden, S & Söderbergh, B 2010, 'The peak of the oil age: analysing the world oil production', *Energy Policy*, vol. 38, no. 3, pp. 1398–1414.

Allison, I, Bindoff, NL, Bindschadler, RA, Cox, PM, de Noblet, N, England, MH, Francis, JE, Gruber, N, Haywood, AM, Karoly, DJ, Kaser, G, Le Quéré, C,

Lenton, TM, Mann, ME, McNeil, BI, Pitman, AJ, Rahmstorf, S, Rignot, E, Schellnhuber, HJ, Schneider, SH, Sherwood, SC, Somerville, RCJ, Steffen, K, Steig, EJ, Visbeck, M & Weaver, AJ 2009, *The Copenhagen Diagnosis: updating the World on the Latest Climate Science*, The University of New South Wales Climate Change Research Centre (CCRC), Sydney.

Anderson, JW 2001, 'How the Kyoto Protocol developed: a brief history', in MA Toman (ed), *Climate change economics and policy: an RFF anthology, resources for the future*, Washington, pp. 11–23.

Aunan, K, Mestl, HE, & Seip, HM 2003, 'Co-benefits of CO_2-reducing policies in China – a matter of scale?' *Global Environmental Issues*, vol. 3, no. 3, pp. 287–304.

BBC 2009, '*Where countries stand on Copenhagen*', viewed 11 may 2017, <http://news.bbc.co.uk/2/hi/science/nature/8345343.stm>.

Bodansky, D 2001, 'The history of the global climate change regime', in U Luterbaoher & DF Sprinz (eds), *International relations and global climate change*, Achorn Graphic Sciences, Inc., Sabon, pp. 23–40.

Chen, G 2012, *China's climate policy*, Routledge, Abingdon.

Earth System Research Laboratory NOAA 2019, *Trends in atmospheric carbon dioxide*, viewed 27 January 2019, <https://www.esrl.noaa.gov/gmd/ccgg/trends/>.

Garrison, JA 2009, *China and the energy equation in Asia: the determinants of policy choice*, First Forum Press, London.

Glomsrød, S & Wei, TY 2016, 'The effects of energy efficiency improvement in China with global interaction', *AIMS Energy*, vol. 4, no. 1, pp. 37–51.

Goldthau, A & Witte, JM 2010, 'The role of rules and institutions in global energy: an introduction', in A Goldthau & JM Witte (eds), *Global energy governance: the new rules of the game*, Brookings Institution, Washington DC, pp. 1–24.

Griffin, JM 2009, *A smart energy policy: an economist's Rx for balancing cheap, clean, and secure energy*, Yale University Press, New Haven.

Guan, D, Hubacek, K, Weber, CL, Peters, GP & Reiner, DM 2008 'The drivers of Chinese CO_2 emissions from 1980 to 2030', *Global Environmental Change*, vol. 18, no. 4, pp. 626–634.

Guan, D, Peters, GP, Weber, CL & Hubacek, K 2009, 'Journey to world top emitter: an analysis of the driving forces of China's recent CO_2 emissions surge', *Geophysical Research Letters*, vol. 36, no. 4, doi:10.1029/2008GL036540.

Heggelund, GM & Buan, IF 2009, 'China in the Asia–Pacific Partnership: consequences for UN climate change mitigation efforts?', *International Environmental Agreements: Politics, Law and Economics*, vol. 9, no. 3, pp. 301–317.

Ho, MS & Nielsen, CP (eds) 2007, *Clearing the air: the health and economic damages of air pollution in China*, MIT Press, Cambridge & London.

Höhne, N, Harnisch, J, Phylipsen, D, Blok, K & Galleguillos, C 2003, *Evolution of commitments under the UNFCCC: involving newly industrialized economies and developing countries*, ECOFYS, Cologne.

Hone, D 2017, *Saving the CDM*, viewed 1 October 2017, <https://blogs.shell.com/2017/05/29/saving-the-cdm/>.

iiEnergy 2016, *China drove 2015 renewables growth*, viewed 20 January 2016, <http://www.ii-energy.com/index.php/2016/01/18/china-drove-2015-renewables-growth/>.

IPCC 2007, 'Summary for policy makers', in ML Parry, OF Canziani, JP Palutikof, PJ ven der Linden & CE Hanson (eds), *Contribution of working group II to the fourth assessment report of the intergovernmental panel on climate change*, Cambridge University Press, Cambridge, pp. 7–22.

Kang, J, Yuan, J, Hu, Z & Xu, Y 2012, 'Review on wind power development and relevant policies in China during the 11th Five-Year-Plan period', *Renewable and Sustainable Energy Reviews,* vol. 16, no. 4, 1907–1915.

Kobayashi, Y 2003, 'Navigating between 'luxury' and 'survival' emissions', in PG Harris (ed), *Global warming and East Asia: the domestic and international politics of climate change,* Routledge, London, pp. 86–108.

Laherrère, J 2004, 'Future of national gas supply', *Proceedings of Conference of Association for the Study of Peak Oil and Gas,* viewed 09 September 2012, <http://www.peakoil.net/JL/JeanL.html>

Le Treut, H, Somerville, R, Cubasch, U, Ding, Y, Mauritzen, C, Mokssit, A, Peterson, T & Prather, M 2007: 'Historical overview of climate change', in S Solomon, D Qin, M Manning, Z Chen, M Marquis, KB Averyt, M Tignor & HL Miller (eds), *Climate change 2007: the physical science basis. Contribution of Working Group I to the Fourth Assessment Report of the Intergovernmental Panel on Climate Change,* Cambridge University Press, Cambridge, pp. 93–112.

Leder, F & Shapiro, JN 2008, 'This time is different: an inevitable decline in world petroleum production will keep oil product prices high, causing military conflicts and shifting wealth and power from democracies to authoritarian regimes', *Energy Policy,* vol. 36, no. 8, pp. 2850–2852.

Lheem, H 2008, 'China's energy security and its strategic implications', in SJ Guo & JMF Blanchard (eds), *"Harmonious World" and China's new foreign policy,* Lexington Books, Lanham, pp. 173–187.

Lin, J 2007, 'Energy conservation investments: a comparison between China and the US', *Energy Policy,* vol. 35, no. 2, pp. 916–924.

Mackerras, C, Taneja, P & Young, G 1998, *China since 1978: reform, modernisation and 'socialism with Chinese characteristics',* 2nd edn, Longman, Melbourne.

Mathews, JA & Tan, H 2014, 'Manufacture renewables to build energy security', *Nature,* vol. 513, no. 7517, pp. 166–168.

McKibbin, WJ & Wilcoxen, PJ 2002, *Climate change policy after Kyoto: blueprint for a realistic approach,* Brookings Institution Press, Washington, D.C.

Mohr, SH & Evans, GM 2009, 'Forecasting coal production until 2100', *Fuel,* vol. 88, no. 11, pp. 2059–2067.

Murray, J & King, D 2012, 'Climate policy: Oil's tipping point has passed', *Nature,* vol. 481, no. 7382, pp. 433–435.

NDRC 2004, *China medium and long term energy conservation plan,* viewed 19 November 2015, <https://policy.asiapacificenergy.org/node/3010>.

NDRC 2014, *China's Policies and Actions on Climate Change 2014,* viewed 03 July 2015, <http://en.ccchina.gov.cn/archiver/ccchinaen/UpFile/Files/Default/2014112613 37277 51798.pdf>.

NDRC 2016, '*News release of NDRC of PRC (English version)*', viewed 05 January 2016, <http://en.ndrc.gov.cn/>.

NEA 2011, *Guowuyuan guanyu yinfa "shi er wu" jieneng jianpai zonghexing gongzuo fang'an de tongzhi [Guofa 26 hao] (State Council announced the notification of the comprehensive work plan for ECERS during the 12th Five-Year Plan [NDRC No. 26 Document]),* viewed 26 May 2017, <http://www.nea.gov.cn/2011-09/08/c_131115016.htm>.

NEA 2015, *'Guojia nengyuanju guanyu xiada 2015 nian dianli hangye taotai luohou channeng mubiao renwu de tongzhi (National Energy Administration's Notification on the target of closure of backward production capacity 2015)', National Energy Administration No. 119 Document,* viewed 20 January 2016, <http://zfxxgk.nea.gov.cn/auto84/201507/t20150728_1951.htm>.

NEA 2016, *12 wu qijiang wei shihua nengyuan xiaofei bizhong tigao (Non-fossil energy consumption proportion increased during the 12th Five-Year Plan)*, viewed 20 January 2016, <http://www.nea.gov.cn/2016-01/05/c_134978582.htm>.

Nyman, E 2015, 'Offshore oil development and maritime conflict in the 20th century: a statistical analysis of international trends', *Energy Research & Social Science*, vol. 6, pp. 1–7.

Obbard, J 2010, 'Climate change – living in the anthropocene', in KL Koh, LH Lye & J Lin (eds), *Crucial issues in climate change and the Kyoto Protocol: Asia and the world*, World Scientific Publishing, Singapore, pp. 63–96.

Oberthür, S & Ott, HE 1999, *The Kyoto Protocol: international climate policy for the 21st century*, Springer-Verlag, Berlin.

Pennock, M, Poland, B & Hancock, T 2016, 'Resource depletion, peak oil, and public health: planning for a slow growth future', in I Luginaah & RB Kerr (eds), *Geographies of health and development*, Routledge, Abington & New York, pp. 177–196.

Qi, Y 2013, *Annual review of low-carbon development in China: 2010*, World Scientific Publishing, Singapore.

Raphael, S & Strokes, D 2016, 'Energy security', in A Collins (ed), *Contemporary security studies*, Oxford University Press, Oxford, pp. 343–355.

Schreurs, MA 2011, 'Climate change politics in an authoritarian state: the ambivalent case of China', in JS Dryzek, RB Norgaard & D Schlosberg (eds), *The Oxford Handbook of Climate Change and Society*, Oxford University Press, Oxford, pp. 449–463.

Stringer, KD 2008, 'Energy security: applying a portfolio approach', *Baltic Security & Defence Review*, vol. 10, no. 1, pp. 121–142.

The Central People's Government of PRC 2006a, *'Wu bumen lianhe fa tongzhi yaoqiu jiaqian zhengfu jigou jieyu ziyuan gongzuo (Five ministries jointly released the Notification on the Reinforcement of Resource Saving within Government Organizations)'*, viewed 15 July 2016, <http://www.gov.cn/jrzg/2006- 03/03/content_216753.htm>.

The World Bank 2004, *China – air, land and water: environmental priorities for a new millennium*, The World Bank Group, Washington DC.

Tian, CR 2016, 'Oil import and export in China', *China Oil and Gas*, vol. 23, no. 1, pp. 36–42.

UNFCCCa n.d, n.p, *'Mechanisms under the Kyoto Protocol'*, viewed 18 March 2013, <https://unfccc.int/process/the-kyoto-protocol/mechanisms>.

Wang, Y 2015, *Taotai luohou channeng heli zhiding guanting xiaohuodian shijianbiao (Eliminating backward production capacity by making sensible schedule for closure of small coal-fired power plants)*, viewed 10 February 2016, <http://info.gongchang.com/t/zhuangbei-797363.html>.

Wei, YM, Liu, LC, Fan, Y & Wu, G 2007, 'The impact of lifestyle on energy use and CO_2 emission: an empirical analysis of China's residents', *Energy Policy*, vol. 35, no. 1, pp. 247–257.

Wen, JB 2005, *The Report on the Work of the Government, delivered by Premier Wen Jiabao at the Third Session of the 10th National People's Congress. March 5, 2005, Beijing*, viewed 04 May 2012, <http://www.chinadaily.com.cn/english/doc/2005-03/15/content_425079.htm>.

Wu, FZ 2013, 'China's pragmatic tactics in international climate change negotiations: reserving principles with compromise', *Asian Survey*, vol. 53, no. 4, pp. 778–800.

Wu, K & Storey, I 2007, 'Energy security in China's capitalist transition', in CA McNally (ed), *China's emergent political economy: capitalism in the dragon's lair*, Routledge, London & New York, pp. 190–208.

Wu, Y & Zhang, W 2016, 'The driving factors behind coal demand in China from 1997 to 2012: an empirical study of input-output structural decomposition analysis', *Energy Policy*, vol. 95, pp. 126–134.

Xinhua News 2015, *Shangbannian qingjie nengyuan tongbi da 17%, kongliang wei shouyao renwu (Proportion of renewable energy reached 17% by mid-2015, the control of total energy consumption and emission should be the primary task)*, viewed 20 January 2016.

Xu, JB 2016, *Zhongguo "shi er wu" jieneng jianpai chengxiao xianzhu, mingnian qidong quanguo tanjiaoyi shichang (China's ECERS achieved significant results during the 12th Five-Year Plan, national carbon trading market is to start next year)*, viewed 26 May 2017, <http://finance.china.com.cn/industry/energy/xnyhb/20160629/3789390.shtml>.

Yao, L & Chang, Y 2014, 'Energy security in China: a quantitative analysis and policy implications', *Energy Policy*, vol. 67, pp. 595–604.

Yergin, D 2006, 'Ensuring energy security', *Foreign Affairs*, vol. 85, no. 2, pp. 69–82.

Yu, HY 2008, *Global warming and China's environmental diplomacy*, Nova Science Publisher, New York.

Zhang, ZH 2004, 'The forces behind China's climate change policy: interests, sovereignty and prestige', in PG Harris (ed), *Global warming and East Asia: the domestic and international politics of climate change*, Routledge, London & New York, pp. 66–85.

Zhang, ZX 2011, 'China's energy security, the Malacca dilemma and responses', *Energy Policy*, vol. 39, no. 12, pp. 7612–7615.

Zhou, Y 2010, 'Why is China going nuclear?', *Energy Policy*, vol. 38, no. 7, pp. 3755–3762.

Zweig, D & Bi, JH 2005, 'China's global hunt for energy', *Foreign Affairs*, vol. 84, no. 5, pp. 25–38.

2 Influences on the development of Chinese policies

Historical and contemporary perspectives

A Confucian governance tradition

Peters (2013) defined public policy as "the sum of government activities, whether pursued directly or through agencies, as those activities have an influence on the lives of citizens" (p. 4). This definition is indeed modern; however, the concept of "citizenship" which comprises the essence of the definition, can be traced back to the direct democracy of ancient Greece (Guarnizo 2017, p. 12; Pocock 1998, p. 33). Although there is no universally agreed-upon definition of citizenship, the development of the notion has almost always occurred around two spheres: membership and exclusion; and rights and obligations (King 2016, p. 9; Zarrow 1997, p. 4). It is this evolution of citizenship throughout Western civilisation that paved the way for modern Western democracies (Wasson 2015, n.p.).

It is the participation of citizens in public affairs, which distinguishes democracies from other systems of government. When switching our lens to the East during the same era and narrowing our focus to China, the concept of citizenship was mentioned by none of the ancient Chinese philosophers, regardless of the fact that the Spring and Autumn and Warring States periods (770 BC–221 BC) were eras when several Chinese classical philosophies flourished. In fact, such a concept has never been proposed by native Chinese scholars until it was introduced from the West in the late 19th century (Jakimow & Barabantseva 2016, p. 168). According to Zarrow (1997, p. 4), citizenship is generally regarded as a notion of the Western civilisation, but has little to do with the East.

In ancient Chinese society, imperial power was seen as paramount. The emperor was regarded as the Son of Heaven, standing above all beings on earth (Link 2015, p. 25). One poem of *The Book of Songs*[1] stated that: "all territories under the sky belong to the emperor, all people who reside on lands and waters are subjects of the emperor". The emperor was granted absolute power by heaven, but he was also expected to fulfil the greatest duty to carry out virtuous ruling and look after his people (Creel 1953, p. 23). The majority of ancient Chinese societies were built on hierarchies and orders based on neo-Confucian principles (theories developed by Confucius' disciples and later followers) known as the *Three Cardinal Guides* and *Five Constant Virtues*.[2] The *Three Cardinal Guides* set up the structure of social order as follows: ruler guides subjects, father guides son, and husband guides wife. The *Five Constant Virtues* regulate interpersonal relations by promoting the qualities of benevolence, righteousness, courtesy, wisdom, and trust (Su 2011, p. 32; Yang 2016, p. 184).

The hierarchy from the top was the emperor, then the class of officials, and the ordinary people at the bottom. The emperor held absolute power; the officials assisted the emperor in managing the ordinary people by providing advice; while the ordinary people looked up to the emperor and officials and wished for their benevolent governance. The ordinary people in ancient Chinese societies were passive and obedient and had little say in public affairs and policies. They were generally content with such status and believed it was the emperor and government's responsibility to implement good policies and guide the empire towards the correct direction. As Liu and Liu (1997) put it, "the consciousness [of being subjects] penetrated deeply into people's hearts and dominated their political behaviour. The rise of a new dynasty was nothing but another cycle in traditional cultural attitudes" (p. 40).

The existence of a class of intellectual officials was another distinct characteristic of Chinese feudal society. The officials were selected from the ordinary people through examinations and worked as a kind of "think tank" for the emperor. This system was considered advanced at the time. The leading figure of Western Enlightenment, Christian Wolff (1679–1754), said in an oration, "in the Art of Governing, this nation [imperial China] has surpassed all others without exception" (Creel 1949, p. 256). The formation of ancient China's intellectual class was based on Confucian principles which held that the government should consist of talented people, and only the knowledgeable and wise could develop and implement good policies for the nation (Bell 2010; Creel 1953; Link 2015). Only if the officials were indeed talented, knowledgeable, and ethical, could they convince and influence the people through setting positive examples, and the ordinary people would look up to the officials, as explained by Confucius:

> If you desire what is good, the people will be good. The moral power of the gentleman is wind, the moral power of the common man is grass. Under the wind, the grass must bend. (*Analects*, 12.19 translated Leys 1997, p. 58)[3]

Therefore, typical policies of imperial China were derived from the works of the sages and the intellectual class. These works were studied by ancient government officials not only as study materials for their examinations, but also as guidance for their duties after they assumed office.

Rules and laws of imperial China were called "*Wang Fa*", or, as the name suggests, the emperor's laws. The emperor's laws were developed by the emperor's think tanks (the intellectual class) and decided by the emperor; the ordinary people had little influence on the legislation (Zhang 2014, p. 450). It is obvious that in imperial China, policy development and decision-making followed a strict hierarchy and worked in a top-down fashion. The ordinary people needed to wait for the wise officials and emperors to spot social problems and then address them. The emperor's vision and the officials' capacity played a very important role in the management of the society and consequently the outcome of people's well-being. In fact, this top-down, highly centralised policy model is applicable to both imperial and modern China. Through the course of interaction between China and the West, a wide range of modern governing concepts and techniques were introduced. Some of these have been adopted such as the separation of the executive, legislative, and judicial powers; however, the Chinese government's policies and activities still maintain unique characteristics resembling this Confucian tradition (Ogden 2005; Wang et al. 2005; Yan 2013). For instance, the Chinese government's reliance on think tanks and intellectuals for public policy development remains heavy. There are various views on the relationship between the Chinese government and its think tanks. Ogden (2005, p. 113) doubts the independence of Chinese think tanks as they rely on the government for funding and promotion, and therefore tend to only work within the boundaries set by the Chinese Communist Party (CCP). Shai and Stone (2004, p. 142) see the reliance on intellectuals as damaging as it "operates within a closed policy context that is distant from civil society". Zhu (2009), on the other hand, argues that the closeness between the Chinese government and its think tanks possesses strengths which the Western system cannot match. For example, how to effectively transfer expert knowledge to policy outcomes and increase the influence of expert knowledge remains a challenge for scholars to tackle in the Western political system. There have been many attempts by Western scholars to build models and micro-mechanisms for solving this problem to ensure evidence-based policy outcomes (Amara, Ouimet & Landry 2004; Lavis et al. 2003; Nutley, Walter & Davies 2007). In China, the transfer of expert knowledge to public policies appears relatively easy compared with that in the West. As Zhu (2009, p. 339) points out, Chinese think tanks are institutionalised and are located within government departments who are directly involved in the drafting of important policies, information release, and initiating policy studies. They are also invited to important government meetings and seminars for immediate policy advice. Furthermore, many think tank scholars are appointed to important decision-maker positions in government departments (Zhu 2013, p. 29). This further enhances the role of the think tank in Chinese policies.

Another distinct characteristic of the Chinese system is the existence of administrative linkages between many semi-official think tanks and the relevant government departments. For example, scholars from public institutions and universities can pass on their research products to their supervising units[4] in the government as a range of alternative considerations for policy decisions (Li 2009, p. 16). Leonard (2008, p. 17) even regards the Chinese intellectuals as an essential force in articulating a broader range of social concerns in a society without opposition parties, independent trade unions, and a free press.

This Confucian tradition of rule by the wise with minimum direct involvement from ordinary people still has great influence on China's modern environmental governance. The constant engagement of intellectuals and experts in China's modern environmental policy remains a characteristic even after the market economy was introduced in the late 1970s. The introduction of market, although diversified Chinese government's policy approaches, could not shake the dominance of the state power.

Policy approaches under China's state neoliberalism

China underwent the transition from "state socialism" to "market socialism" after the 1978 reform and opening up. The introduction of the market economy inevitably involved deregulation, marketisation, and privatisation, which presented neoliberal capitalist characteristics. However, the communist party state firmly held its guiding ideology – socialism, and its political and administrative power could not be challenged. The term "state neoliberalism" or "state capitalism" is used to describe Chinas situation: a socialist state that led national development along neoliberal capitalism path (So & Chu 2016, p. 56). Under such a system, it has seen dramatic diversification of policy approaches in the public policy domain, especially in environmental policies. The administrative approach, although remains dominant, is gradually accompanied by market-based policy approach. The voice calling for democratic approach is also emerging for solving local problems.

The dominance of administrative rationalism in China's environmental governance

Dryzek (2013) defines administrative rationalism as a "problem-solving discourse which emphasises the role of the expert rather than the citizen or producer/consumer in social problem solving, and which stresses social relationships of hierarchy rather than equality or competition" (p. 75). As a problem-solving discourse, administrative rationalism is not about public participation, but about "rational management in the service of a clearly defined public interest, informed by the best available expertise" (Dryzek 2013, p. 88). This is to say, public interest, in such an approach, is uncovered and conceptualised by experts through technical procedures such as professional policy analysis rather than articulated and raised by the general public.

There are a range of institutions and practices involved in administrative rationalism when dealing with environmental problems. These often include bureaucracies of national resource management, pollution control agencies, regulatory policy instruments, environmental impact assessment, expert advisory commissions, planning, and rationalistic policy analysis techniques (Dryzek 2013, pp. 76–84). A top-down approach to planning is the most predominant characteristic of China's governance of its environmental problems. The central government usually sets up a national goal and then divides and appoints the tasks for provincial and municipal governments (Lo 2014, pp. 238–239) such as those in its Energy Conservation and Emission Reduction Scheme (ECERS) during the 11th and 12th Five-Year Plans. Rationalistic policy analysis, such as cost and benefit analysis, is widely performed by the experts in the policy process. However, the national resource management bureaucracies, pollution control agencies, environmental impacts assessment, and regulatory policy instruments are weak in China (Carter & Mol 2013). This situation is caused by China's previous pursuit of economic development at all cost. Nonetheless, this situation is gradually changing as China started to realise the importance of resource conservation and pollution control on the way to a harmonious society (Zheng & Tok 2007). As a result, the status of national bureaucracies for resource conservation and pollution controls has gradually improved. The newly established Ministry of Natural Resources (MNR) and Ministry of Ecology and Environment (MEE) are two of the representatives of such improvement.

Bureaucracy is essential in a system of administrative rationalism. The 19th-century German sociologist Max Weber's theory of bureaucracy provided foundation for the management of environmental problems under administrative rationalism. It was believed that isolated individuals were unable to solve increasingly complicated social and economic problems; therefore, coordinated solutions were required. Bureaucracy provided the framework for this to be achieved, in which complex problems could be divided and then subdivided into portions small enough for assigned individuals or small groups to work out a solution (Weber 2009). Such a system has advantages when implementing large-scale programs, for instance, China's renewable energy development program. The central government sets up the national goal for renewable energy expansion within a five-year time frame, and then divides the tasks and distributes them to each province. Each province further divides and appoints the tasks to municipal governments.

The idea is simple and the implementation is reasonably straightforward. However, such a system also exhibits weaknesses, such as the quality of implementation. Under administrative rationalism, monitoring and auditing are especially important procedures for all levels of the administrative hierarchy. However, such procedures are often lacking at lower levels of government where the central government's influence is weaker. As a result, the quality of the policy implementation at the local level is sometimes poor. The gap between what the government aims to achieve and what is actually achieved at the local level is often termed as the "implementation deficit" (McLaughlin & Krantzberg 2011; Mitchell 2014; Weale 1992). Developers and polluters under a municipal or a county government's administration are often unwilling to comply with the central government's policy decision as this causes the reduction of profits (Dryzek 2013, p. 95). This is exactly the situation in China's environmental governance (Zhang & Cao 2015). Monitoring and auditing of such behaviour within the vast scope of the lower end administrative system would require a large amount of resources, which is, in most cases, lacking. It is obvious that a policy delivered through the administrative approach can be expensive. According to Ye (2013), China's energy saving and emission production program delivered though the administrative approach was five times more expensive than Clean Development Mechanism (CDM). China understands that the administrative approach is not optimal to achieve efficient resource allocation and utilisation. It, therefore, actively experimented with market mechanism in its climate-related policies.

Market mechanisms in China's climate change mitigation

The Chinese government actively explores market options to solve its environmental problems. China's top leaders have long realised the power of the market. The nation's economic growth since its introduction of a market economy also proved this (Morrison 2009, pp. 4–6). On the management of climate change, the Chinese government experimented with two market-based mechanisms, namely the CDM introduced in the early 2000s, and the pilot emission trading scheme (ETS), introduced in 2012. The national ETS commenced at the end of 2017 (China Environmental Protection Online 2018). Stavins (2003) defines market-based instruments as follows:

> Market-based instruments are regulations that encourage behaviour through market signals rather than through explicit directives regarding pollution control levels or methods. These policy instruments, such as tradable permits or pollution charges, are often described as "harnessing market forces" because if they are well designed and implemented, they encourage firms (and/or individuals) to undertake pollution control efforts that are in their own interests and that collectively meet policy goals. (p. 298)

According to Stavins (2008, p. 300), traditional administrative instruments tend to set up uniform technological and performance-based standards for firms and distribute shares of pollution control duties to them to accomplish regardless of cost. This may end up being very expensive because different firms may require different technologies in controlling their emissions. Under the uniform standard, firms may spend more on the required pollution control procedure rather than the most cost-effective procedures specially catered for their own situations. Market-based instruments, in contrast, provide flexibility. In theory, the maximum environmental benefit can be achieved with minimum cost through intelligently designed and implemented market-based instruments (Lockie 2013). This is because market-based instruments provide incentives for firms to adopt the most cost-efficient abatement options to achieve the optimal emission reduction outcome rather than following the set standard method (Baumol & Oates 1988; Tietenberg 1995).

Flexibility and cost-efficiency are the predominant advantages of market-based instruments; however, they are not without flaws in reality. Neuteleers and Engelen (2015, p. 253) argued that there should be good reasons to question the integrity of market-based monetary valuation in the environmental domain because it could be morally problematic in the commodification process of environmental goods. Spash and Aslaksen (2015) stressed the challenges of traditional environmental economics in the valuation of the ecosystem and biodiversity. Kroeger and Casey (2007, p. 321) also pointed out that in reality markets fail in delivering efficient allocation of ecosystem services on many occasions. In such cases, well-designed government interventions will be required to protect the public interest. Dryzek (2013, p. 130) indicated that the tumbling emission permit price and the broken down EU ETS after the hit of the 2008 global financial crisis were due to the initial design flaw of the scheme, where it did not set up a stringent enough sum of emission permits. The initial over-allocation of permits was a very significant trigger of the collapse of the permit price. Therefore, it is possible for market mechanisms to fail.

China's market-based instruments in the management of climate change have some special characteristics. First of all, these market mechanisms are still heavily influenced by China's administrative power. As Lo and Howes (2014) pointed out, "The Chinese carbon markets are characterised by a hierarchical relationship between the regulator and many regulated enterprises" (p. 397). According to Kossoy and Guigon (2012, p. 98), China's Department of Price may play an important role in deciding the price of carbon and managing its fluctuation in the context of a regulated power market if the power sector is to be included in the national trading scheme. Regulatory authorities contributed greatly to the good compliance results in the pilot carbon trading programs, as failure in compliance means losing face for the responsible authorities (Lo & Howes 2015).

The second characteristic of China's climate change market instruments is that they are economic and security-driven rather than emission mitigation driven. Lo (2013, p. 73) saw China's carbon trading as an incentive for corporates to carry out low-carbon economic development, but carbon emission reduction was only a co-product. It was regarded as primarily an economic policy providing some environmental benefits instead of an environmental policy initiated by an environmental agency. This was due to the heavy involvement of the National Development and Reform Commission (NDRC), an economic planning commission, in the initial policy process. Lewis (2010) argued China's active participation in international CDM was due to its concerns for the negative impacts of climate change on its economically advanced regions in the eastern part of the country. Its participation was also for the purpose of attracting international finance for its renewable energy development. These characteristics indicate that market mechanisms, even as a policy approach that China embraces, have a hue of administrative colour.

Regardless the initial intention, the market approach has played an important role in China's climate-related energy policies. Since the initial participation in CDM program, China has become the biggest CDM hosting country. China's national ETS is so far the world's largest ETS. It initially covers over 1700 enterprises in China's electricity industry. Approximately 3.5 billion tons of carbon emission are traded in the national carbon market, taking up 39% of national total (China Environmental Protection Online 2018). The market approach is an effective addition in China's climate change governance. It is expected to largely improve the efficiency and cost-effectiveness in China's energy saving and emission reduction programs.

Apart from the administrative and market approaches, the democratic approach is generally believed to have a positive impact on the control of pollution in the developed world (Farzin & Bond 2006; Li & Reuveny 2006). For developing countries, Mak Arvin and Lew (2011) also argue that democracy is conducive to improving environmental quality. For China, although some decentralised initiatives have been carried out in local economic management and village elections (Epstein 1997, p. 406; Landry 2008, p. 221; Wang & Yao 2007), using a democratic approach to solve local environmental problems directly is still not common. Nonetheless, optimistic signs started to emerge in China's environmental policy process. New actors beyond the government and large State-owned enterprises (SOEs) expressed their intention to be involved in policy development and revision. China's public policy framework is no longer a tightly closed up system.

An updated fragmented authoritarianism framework

The fragmented authoritarianism framework was first highlighted by Lieberthal and Oksenberg (1988) in their study of the politics, bureaucratic structure, and process of China's energy policies. The authors argued that understanding of the bureaucratic structure and policy process was of great importance in the analysis of Chinese politics and public policies. They argued that "the complex structure of the Chinese state itself as a significant determinant of political process and policy outcomes" should not be overlooked (Lieberthal & Oksenberg 1988, p. 3). Lieberthal and Oksenberg's analysis showed that fragmentation was a significant characteristic of China's energy sector and the Chinese bureaucratic system as a whole. This fragmented authoritarianism tended to encourage competition among ministries with the same level of authority. Among the top leaders and the central ministries, the politics were played based on hierarchy. Ministries, even those on the same level of the hierarchy, may have different levels of influence on policy decisions based on the ranking of top leaders behind them. The higher ranking officials had the power to raise their policy preferences among competing options. Therefore, the agenda setting and policy decisions at the centre were a reflection of the power, status, and views of the top leaders (Lieberthal & Oksenberg 1988, p. 23). On the other hand, although the top leaders were very efficient and effective at commanding the central bureaucracies, the implementation outcomes at the provincial and local levels could vary dramatically from Beijing's pre-set policy goals. This was because the policy implementation process also involved multi-bureaucratic bargains, negotiations, and compromises as a means for overcoming the fragmentation. For instance, a province believed to be of more economic significance by the top leaders can get more resources. However, if a less significant province wants to obtain more resources, the provincial officials have to seek to build up connections with the central leaders (Liebertal & Oksenberg 1988, p. 32). Lieberthal and Oksenberg's fragmented authoritarian theory set up an important framework for the study of Chinese politics and public policies. Several decades have passed since the proposal of Lieberthal and Oksenberg's theory. There have been changes and new insights in China's politics, bureaucratic structures, and policy processes. However, the basic characteristic of fragmentation within the Chinese authoritarian system remains unchanged.

Gilli, Li and Qian (2016) identified this continued fragmentation in China's social welfare sector. Their analysis concluded that the top-down policy process combined with serious competition among vertical bureaucracies for resources had resulted in inefficiency in resource allocation and insufficiency in infrastructure input. Marks (2010, p. 972) also confirmed the continuity of fragmented authoritarianism in China's recent climate change policy processes. He noted that extensive negotiations and bargaining happened both among central bureaucratic agencies during the policy formation process and at all levels of bureaucratic agencies during the implementation process. The more powerful regions and bureaucracies had the power to shape the policy towards their preferences. This lengthy negotiation process twisted the policy into different outcomes when it finally reached down to different localities (Marks 2010, p. 976).

Nonetheless, things are improving. Although the fragmented authoritarian system is prone to problematic implementation due to, for instance, bureaucratic competition, the pro-development local officials, and weak legal and environmental state (Marks 2010; Shirk 2014; van Rooij, Stern & Fürst 2016), there has been a rapid growth in China's renewable energy sector, which are the result of a set of more logical and coherent policies. Lema & Ruby (2007) analysed the policy change in China's wind energy sector during a 20-year (1986–2006) time period. They concluded that this process reflected the policy transformation from fragmentation to coordination in China's wind energy sector. In Lema & Ruby's analysis, the central government's pragmatic approach towards energy was the main driver for the policy changes in the wind power sector.

This echoed one of Lieberthal and Oksenberg's (1988, p. 3) hypotheses that Chinese policy change was due to the top leaders' pragmatic reaction to an emerging or urgent issue. In the 1980s, the power shortage which had impacted one fifth of the nation's industries forced the state to abandon its monopoly of the energy sector by allowing private investments into small-scale power plants (Lema & Ruby 2007, p. 3880). In the 1990s, the top decision-makers realised the harmful consequences of coal-burning on human health and the environment. This new consciousness led to China's participation in multiple international environmental treaties, including the Kyoto Protocol. This process paved the way for a slightly improved regulatory capacity of the renewable energy sector. However, problems such as coordinating difficulty, unstable demand of wind power, and investment risks could not be overcome during the implementation because provincial grids could not be convinced of the profit of wind power, and therefore were unwilling to connect wind power to the grid (Lema & Ruby 2007, p. 3882).

In response to these problems, the central government adopted two procedures in the early 2000s, including centralising the energy bureaucracy and decentralising the market. The multiple competing and overlapping bureaucratic agencies involved in the renewable energy sectors were merged into one renewable energy department under the National Energy Administration of NDRC, which was to say, renewable energy policy was governed by only one agency with the ability to produce coherent policies since then. In terms of market decentralisation, energy production and transmission were separated. Two grid SOEs, the National Grid Corporation of China and China Southern Grid Corporation were established. The original China State Power Corporate was divided into five energy SOEs, Huangneng, Huandian, China Datang, Guangdian, and China Investment Group, in order to create competition. The more coordinated central policies and increased competition led to fast development of China's wind energy in the early and mid-2000s (Lema & Ruby 2007, p. 3882). However, since the late 2000s, this rapid increase of wind power capacity started to present new problems, such as the wastage of wind power, which will be explored later in this book.

The transformation of China's renewable energy policy from fragmentation to coherence demonstrates that fragmented authoritarianism can be reformed. However, it required the vision of the top decision-makers (Lieberthal & Oksenberg 1988, p. 34). Any unwise decisions could lead to completely different consequences. Therefore, consideration of the outside opinions from the non-state stakeholders such as private investors, NGOs and the general public would be a step forward for China's policy development and implementation.

Three decades after Lieberthal and Oksenberg's proposal of the fragmented authoritarianism framework, new perspectives have emerged on China's fragmented authoritarian style policy development. Mertha (2009, p. 996) identified that the Chinese policy process embedded in the fragmented authoritarian system started to provide opportunity for non-traditional policy actors such as NGOs, journalists, and even influential individuals to enter the policy process, providing fertile ground for policy changes. In Mertha's (2009) words, these newly emerged policy actors "adopt the strategies [bargains, negotiations, pressuring, etc.] that traditional actors in China have used for decades to pursue their agendas and institutional mandates" (p. 997).

Among the new policy actors, Mertha (2009, p. 997) identified the tight connection between the media and the NGOs, due to the fact that many NGO leaders were trained as journalists or editors. The most common way the media and NGOs recruited support was through the articulation of convincing narratives from an unconventional angle, which opened up debates for policy issues. This issue framing process is important as it delivers critical stories in a way that the general public can understand and are interested in, which in turn accumulates support and enhances the bargaining power in the policy negotiation process. The issue framing is also a process for the new actors to test out the boundaries of the authoritarian government. In reality, the new actors tend to work strategically in the policy process. Rather than completely opposing a policy, they carefully work within the political boundary by seeking only partial modification to certain parts of the policy in order to manoeuvre the policy outcome closer to their goal. Mertha (2008, p. 151) termed such actions as "a kinder and gentler fragmented authoritarianism", because the new policy actors "do not threaten – and are not seen as threatening – the legitimacy of the government or the Chinese Communist Party". Other scholars termed the allowance of non-state actors' involvement as "responsive authoritarianism", which was regarded as an improvement in China's policy processes towards a more plural and inclusive governance (He & Warren 2011; Reilly 2011; Weller 2008).

China's current bureaucratic system and policy priorities are obviously different from that when Lieberthal and Oksenberg outlined the fragmented authoritarianism framework. However, ministerial bargaining and fragmentation still persist, as pointed out by Grünberg (2017) in China's current energy sector. Therefore, the fragmented authoritarianism model "has not altogether lost its value as an analytical framework to describe the important structural features and operational mechanisms of China's political system" (Grünberg 2017, p. 16). The main structure of the Chinese system remains unchanged. In Lieberthal and Oksenberg's (1988) original hypothesis, describing the structural fragmentation of the Chinese authoritarian system, they stated:

> Policy X resulted from a bargain among Ministries A, B, and C and Province D … Disgruntled Ministries E and F, losers in the deal, planned to pursue strategies to erode the agreement. The bargain sought to reconcile the conflicting organisational missions, ethos, structure, and resource allocations of the ministries involved. (pp. 3–4)

This hypothesis is still applicable to contemporary China's bureaucratic structure and policy process. However, there should be a new element added to it, namely the emerging but limited input of new policy actors in the fragmented authoritarian system. In later chapters (3, 4, and 5), this updated version of fragmented authoritarianism will be used as an analytical framework for China's climate-related energy actors and policies.

In sum, this chapter set out the conceptual framework of this book by exploring China's policy style, approaches, and framework from both historical and developing angles. It identified the influence of classic Confucian principle of "rule by the wise" on China's consistent authoritarian style policies throughout Chinese history. It also demonstrated how the administrative and market approaches were applied to climate-related policies in China.

In particular, it explored the ongoing relevance of the fragmented authoritarianism framework as a means of analysing such policies from their design to implementation, and proposed an extended version of this framework to incorporate other actors beyond the government. In doing so, this chapter prepared the ground in later chapters for the analyses of China's climate-related energy actors, and their roles in CDM and ECERS.

Notes

1. *The Book of Songs,* formerly called *Poetry,* is a collection of 305 ancient Chinese poems from the early Western Zhou Dynasty (the 11th century BC) to the mid-Spring and Autumn Period. It is said that they had been compiled by Confucius. *The Book of Songs* contains three sections based on its contents, namely *Feng* (*ballads*), *Ya* (*poems from intellectuals or aristocrats*), and *Song* (*songs for praying*). The poem mentioned here is from *Xiaoya. Gufengzhishi, Beishan.*
2. Western Han Confucian scholar Dong Zhongshu modified Confucius' thought on social hierarchy and status and theorized the *Three Bonds/Cardinal Guides* and *Five Constant Virtues* in his work *Chun Qiu Fan Lu.*
3. For Confucius, a gentleman is someone who is "benevolent, wise, and reverent" (Van Norden 2011, p. 20). Some scholars use the term exemplary person instead of gentleman (see, e.g. Bell 2010).
4. The majority of the famous Chinese universities, such as those listed in China's "211 Project" and "985 Project", are state supervised, and therefore have their supervising units within the government (Ministry of Education of PRC 2015).

References

Amara, N, Ouimet, M & Landry, R 2004, 'New evidence on instrumental, conceptual, and symbolic utilization of university research in government agencies', *Science Communication*, vol. 26, no. 1, pp. 75–106.

Baumol, WJ & Oates WE 1988, *The theory of environmental policy*, 2nd edn, Cambridge University Press, New York.

Bell, DA 2010, *China's new Confucianism: Politics and everyday life in a changing society*, Princeton University Press, Princeton, NJ.

Bell, DA 2010, *China's new Confucianism: Politics and everyday life in a changing society*, Princeton University Press, Princeton.

Carter, N & Mol, AP (eds) 2013, *Environmental governance in China*, Routledge, London & New York.

China Environmental Protection Online 2018, *Quanguo tanjiaoyi banniankao: duochong jiji xinhao zhicheng shichang wending xianghao (National ETS half-year assessment: multiple signals indicating positive market)*, viewed 30 January 2019, <m.hbzhan.com/News/Detail/124759.html?=singlemessage&isappinstalled=0>.

Creel, HG 1949, *Confucius and the Chinese way*, Harper Torchbooks, New York.

Creel, HG 1953, *Chinese thought from Confucius to Mao Tse-tung*, The University of Chicago Press, London.

Dryzek, JS 2013, *The politics of the earth: environmental discourses*, Oxford University Press, Oxford.

Epstein, AB 1997, 'Village elections in China: experimenting with democracy', in Joint Economic Committee of the US Congress (ed), *China's economic future: challenges to US policy*, ME Sharpe, Armonk, pp. 403–22.

Farzin, YH & Bond, CA 2006, 'Democracy and environmental quality', *Journal of Development Economics*, vol. 81, no. 1, pp. 213–235.

Gilli, M, Li, Y & Qian, J 2016, *Logrolling under fragmented authoritarianism: theory and evidence from China*, Dems Working Paper Series, No. 333, viewed 8 August 2017, <http://dems.unimib.it/repec/pdf/mibwpaper333.pdf>.

Grünberg, N 2017, 'Revisiting fragmented authority in China's central energy administration', in KE Brosdgaard (ed), *Chinese politics as fragmented authoritarianism: earthquakes, energy and environment*, Routledge, London & New York, pp. 15–37.

Guarnizo, LE 2017, 'The fluid, multi-scalar and contradictory construction of citizenship', in MP Smith & M McQuarrie (eds), *Remaking urban citizenship: organizations, institutions and the right to the city*, Routledge, London & New York, pp. 11–38.

He, B & Warren, ME 2011, 'Authoritarian deliberation: the deliberative turn in Chinese political development', *Perspectives on Politics*, vol. 9, no. 2, pp. 269–289.

Jakimow, MJ & Barabantseva, EV 2016, '"Othering" in the construction of Chinese citizenship', in L Koenig & B Chaudhuri (eds), *The politics of the 'other': western concepts in non-western contexts – case studies from India and China*, Routledge, Abingdon, pp. 167–178.

King, DS 2016, 'Citizenship as obligation in the United States: title II of the Family Support Act of 1998', in U Vogel & M Moran (eds), *The frontiers of citizenship*, Springer, New York, pp. 1–31.

Kossoy, A & Guigon, P 2012, *State and trends of the carbon market 2012*, The World Bank, Washington, DC.

Kroeger, T & Casey, F 2007, 'An assessment of market-based approaches to providing ecosystem services on agricultural lands', *Ecological Economics*, vol. 64, no. 2, pp. 321–332.

Landry, PF 2008, *Decentralised authoritarianism in China: the Communist Party's control of local elites in the post-Mao Era*, Cambridge University Press, New York.

Lavis, JN, Robertson, D, Woodside, JM, McLeod, CB & Abelson, J 2003, 'How can research organizations more effectively transfer research knowledge to decision makers?', *The Milbank Quarterly*, vol. 81, no. 2, pp. 221–248.

Lema, A & Ruby, K 2007, 'Between fragmented authoritarianism and policy coordination: creating a Chinese market for wind energy', *Energy Policy*, vol. 35, no. 7, pp. 3879–3890.

Leonard, M 2008, *What does China think?*, Fourth Estate, London.

Lewis, JI 2010, 'The evolving role of carbon finance in promoting renewable energy development in China', *Energy Policy*, vol. 38, no. 6, pp. 2875–2886.

Leys, S 1997, *The Analects of Confucius*, W.W. Norton & Company, New York.

Li, C 2009, 'China's new think tanks: where officials, entrepreneurs, and scholars interact', *China Leadership Monitor*, no. 29, pp. 1–21. <https://doi.org/10.1142/9789813100244_0006>.

Li, Q & Reuveny, R 2006, 'Democracy and environmental degradation', *International Studies Quarterly*, vol. 50, no. 4, pp. 935–956.

Lieberthal, K & Oksenberg, M 1988, *Policy making in China: leaders, structures, and processes*, Princeton University Press, Princeton, NJ.

Link, P 2015, 'What it means to be Chinese', *Foreign Affairs*, vol. 94, no. 3, pp. 25–31.

Liu, ZH & Liu, JQ 1997, 'Civic associations, political parties, and the cultivation of citizenship consciousness in modern China', in JA Fogel & PG Zarrow (eds), *Imagining the people: Chinese intellectuals and the concept of citizenship, 1890-1920*, ME Sharpe, New York, pp. 39–60.

Lo, AY 2013, 'Carbon trading in a socialist market economy: can China make a difference?', *Ecological Economics*, vol. 87, pp. 72–74.

Lo, AY & Howes, M 2015, 'Power and carbon sovereignty in a non-traditional capitalist state: discourses of carbon trading in China', *Global Environmental Politics*, vol. 15, no. 1, pp. 60–82.

Lo, K 2014, 'China's low-carbon city initiatives: the implementation gap and the limits of the target responsibility system', *Habitat International*, vol. 42, pp. 236–244.

Lockie, S 2013, 'Market instruments, ecosystem services, and property rights: assumptions and conditions for sustained social and ecological benefits', *Land Use Policy*, vol. 31, pp. 90–98.

Mak Arvin, B & Lew, B 2011, 'Does democracy affect environmental quality in developing countries?', *Applied Economics*, vol. 43, no. 9, pp. 1151–1160.

Marks, D 2010, 'China's climate change policy process: improved but still weak and fragmented', *Journal of Contemporary China*, vol. 19, no. 67, pp. 971–986.

McLaughlin, C & Krantzberg, G 2011, 'An appraisal of policy implementation deficits in the Great Lakes', *Journal of Great Lakes Research*, vol. 37, no. 2, pp. 390–396.

Mertha, A 2008, *China's water warriors: citizen action and policy change*, Cornell University Press, Ithaca & London.

Mertha, A 2009, '"Fragmented authoritarianism 2.0": political pluralization in the Chinese policy process', *The China Quarterly*, vol. 200, pp. 995–1012.

Ministry of Education of PRC 2015, ' *"211 Gongcheng" yu "985 Gongcheng" ("211 Project" and "985 Project")*', viewed 1 October 2017, <http://www.moe.edu.cn/jyb_xwfb/xw_zt/moe_357/jyzt_2015nztzl/2015_zt15/15zt15 _mtbd/201511/t20151106_217950.html>.

Mitchell, B 2014, 'Addressing implementation deficits related to IWRM in Canada', in VR Squires, HM Milner & KA Daniell (eds), *River basin management in the twenty-first century: understanding people and place*, CRC Press, Boca Raton, pp. 42–59.

Morrison, WM 2009, *China's economic conditions*, viewed 11 May 2017, <file:///D:/PhD%20Thesis/Draft%202/China's%20econoimc%20condition.pdf>.

Neuteleers, S & Engelen, B 2015, 'Talking money: how market-based valuation can undermine environmental protection', *Ecological Economics*, vol. 117, pp. 253–260.

Nutley, SM, Walter, I & Davies, HT 2007, *Using evidence: how research can inform public services*. Policy Press, Bristol.

Ogden, S 2005, 'The changing relationship of Chinese intellectuals with the party-state', in M Goldman & E Gu (eds), *Chinese intellectuals between state and market*, Routledge, London, pp. 111–137.

Peters, BG 2013, *American public policy: promise and performance*, 9th edn, Sage, Thousand Oaks, CA.

Pocock, JGA 1998, 'The ideal of citizenship since classical times', in G Shafir (ed), *The citizenship debates: a reader*, University of Minnesota Press, Minneapolis, MN, pp. 33–55.

Reilly, J 2011, *Strong society, smart state: the rise of public opinion in China's Japan policy*, Columbian University Press, New York.

Shai, MC & Stone, D 2004, 'The policy tradition of Chinese policy research institutes', in D Stone & A Denham (eds), *Think tank traditions: policy research and politics of ideas*, Manchester University Press, Manchester, pp. 141–162.

Shirk, S 2014, 'The domestic context of Chinese foreign security policies', in S Pekkanen, C Ravenhill & R Foot (eds), *The Oxford handbook of the international relations of Asian*, Oxford University Press, New York, pp. 391–410.

So AY & Chu YW 2016, *The global rise of China*, Polity Press, Cambridge & Malden, MA.

Spash, CL & Aslaksen, I 2015, 'Re-establishing an ecological discourse in the policy debate over how to value ecosystems and biodiversity', *Journal of Environmental Management*, vol. 159, pp. 245–253.

Stavins, RN 2003, 'Experience with market-based environmental policy instruments', *Handbook of environmental economics*, Nota di Lavoro, Fondazione Eni Enrico Mattei, no. 52.2002, pp. 355–435.

Stavins, RN 2008, 'Addressing climate change with a comprehensive US cap-and- trade system', *Oxford Review of Economic Policy*, vol. 24, no. 2, pp. 298–321.

Su, J 2011, 'On cultural function from Dong Zhong-Xu's idea of "three cardinal guides and five constant virtues"', *Journal of Hunan City University*, vol. 32, no. 6, pp. 32–34.

Tietenberg, T 1995, 'Tradeable permits for pollution control when emission location matters: what have we learned?', *Environmental and Resource Economics*, vol. 5, pp. 95–113.

Van Norden, B 2011, *Introduction to classical Chinese philosophy*, Hackett Publishing Company, Cambridge.

van Rooij, B, Stern, RE & Fürst, K 2016, 'The authoritarian logic of regulatory pluralism: understanding China's new environmental actors', *Regulation & Governance*, vol. 10, no. 1, pp. 3–13.

Wang, J, Wang, GG, Ruona, WE & Rojewski, JW 2005, 'Confucian values and the implications for international HRD', *Human Resource Development International*, vol. 8, no. 3, pp. 311–326.

Wang, S & Yao, Y 2007, 'Grassroots democracy and local governance: evidence from rural China', *World Development*, vol. 35, no. 10, pp. 1635–1649.

Wasson, DL 2015, 'Roman Government', *Ancient History Encyclopedia*, viewed 1 June 2016, <http://www.ancient.eu/Roman_Government>.

Weale, A 1992, *The new politics of pollution*, Manchester Uniersity Press, Manchester.

Weber, M 2009, *The theory of social and economic organization*, AM Henderson & T Parsons (trans.), The Free Press, New York.

Weller, RP 2008, 'Regime responses, responsive authoritarianism', in B Gilley, L Diamond & W Chen (eds), *Political change in China; comparisons with Taiwan*, Lynne Rienner Publishers, Boulder, CO, pp. 117–138.

Yan, XT 2013, *Ancient Chinese thoughts, modern Chinese power*, Princeton University Press, viewed 30 May 2017, <https://muse.jhu.edu/book/46443>.

Yang, CL 2016, 'Chinese philosophy, values, and culture', in CT Foo (ed), *Diversity of managerial perspectives from inside China*, Springer, Singapore, pp. 181–193.

Zarrow, PG 1997, 'Citizenship in China and the West', in JA Fogel & PG Zarrow (eds), *Imagining the people: Chinese intellectuals and the concept of citizenship, 1890-1920*, ME Sharpe, New York, pp. 39–60.

Zhang, B & Cao, C 2015, 'Four gaps in China's new environmental law: implementation and accountability will remain challenging, especially at the local level', *Nature*, vol. 517, no. 7535, pp. 433–435.

Zhang, JF 2014, *The tradition and modern transition of Chinese law*, Springer Science & Business Media, Berlin.

Zheng, Y & Tok, SK 2007, 'Harmonious society and harmonious world: China's policy discourse under Hu Jintao', Briefing Series, no. 26.

Zhu, XF 2009, 'The influence of think tanks in the Chinese policy process: different ways and mechanisms', *Asian Survey*, vol. 49, no. 2, pp. 333–357.

Zhu, XF 2013, *The rise of think tanks in China*. Routledge, Abington.

3 Actors in China's climate-related energy policies

Introduction

Policy analysis is multidimensional which can be either content-focused or actor-focused. Walt and Gilson (1994, p. 353) argued for the importance of actor-focused analysis. This is because actors, or stakeholders, play an important role in policy formation and implementation. The analysis of actors helps to answer the question of how and why policies succeed or fail to deliver desired outcomes. Climate change is of great complexity, which interconnects multiple actors on numerous social and environmental issues such as air pollution, energy structure, transport policy, renewable energy, and carbon sinks (Dryzek 2013, p. 9). Therefore, the analysis of relevant actors, the interactions between actors, and the impacts of such interactions are important for the understanding of why and how climate-related policies are developed and implemented. China, although managed under an authoritarian government, has seen an increasing number of actors involved in its climate-related policies (Williams 2014, p. 1). This chapter identifies four major actors in China's climate-related policies, namely the relevant government departments, the think tanks, enterprises, and the civil society. It examines the behaviour and power of these actors and their influence on China's climate-related energy policies. Through this analysis, it is anticipated to gain an understanding of which actor plays the most significant role in China's climate-related energy policy formation and implementation processes, and what this status means for the pursuit of better climate-related energy policies.

Decentralisation and actors in China's climate change affairs

China has undergone nationwide reform after 1978. The opening up to the outside world allowed China to work with a range of international organisations. This process encouraged transparency in Chinese decision-making processes to some extent. Interest groups and individuals outside the Chinese government also increased in number, with the purpose of having their say and influence on Chinese public policies (Sutter 2008, p. 39). Overall, reform led to a decentralisation process in both administrative and fiscal terms (Christensen, Dong & Painter 2008, p. 352; Montinola, Qian & Weingast 1995; Xu 2011, pp. 1076–1080). Although the Chinese Communist Party (CCP) and the central government continued to hold absolute power in governing the nation (Landry 2008; Tsui & Wang 2008; Yang 2006), the provincial and municipal governments were granted more power and resources for local affairs (Williams 2014; Xu 2011). Continuous administrative reform also allowed more discussions and debates among central ministries on a range of social and economic affairs (Christensen, Dong & Painter 2008). According to Sutter (2008, p. 39), fiscal decentralisation increased the proportion of privately owned enterprises, and enabled state-owned enterprises (SOEs) to compete in the market. As China's economy progressed, the business sector gained increasing influence on a range of economic and social policies (Shirk 1993, p. 16). Other social actors, such as the expert community, the media, the environmental non-governmental organisations (ENGOs) and the general public, also grew to various extents against a background of reform and opening up (Richerzhagen & Scholz 2008). The diversification of actors was also seen in climate change affairs, where they sought to influence the policy direction and to shape the policy outcomes.

China's central government was the main actor in forming climate-related policies (Richerzhagen & Scholz 2008, p. 318) during the investigated period. Provincial and local governments played an important role in the implementation of the central government's climate-related policies. The think tank community, or the experts, gained increasing influence on China's climate-related policies due to the relatively new status of climate affairs and the lack of expertise within the government. Large energy SOEs played an important role in the efficiency improvement of coal-fired power plants, and therefore had great capacity in China's climate change mitigation (Bergsager & Korppoo 2013, pp. 40–41). Private investors, especially those that had invested in renewable energies, embraced the central government's decisions in tackling climate change and welcoming a range of incentive-driven policies. The civil society wished to express their concerns, but their actions faced great challenges. Therefore, it played an increasing but restrained role in the making of China's climate-related policies (Richerzhagen & Scholz 2008, p. 318).

Government ministries

The investigation into the politics of climate change in China required a hierarchical approach (Qi & Wu 2013, p. 301). Understanding the dynamics within the central government and the interactions between the central leadership and local authorities were the keys to comprehending how national climate change policies were developed and implemented. This section highlights these two sets of dynamics within China's fragmented authoritarianism, the horizontal relationship within the central ministries and the vertical relationships between central, provincial and local governments. It does so through an analysis of major ministries involved in climate change affairs and their relations with provincial and local governments.

The central government of China held responsibility for decision-making on climate-related policies which was the most powerful actor in China's climate affairs. At the top, China's National Leading Group on Climate Change was established on 12 June 2007 by the State Council, the chief administrative authority of China. The National Leading Group aimed at coordinating the complex decision-making process of climate-related policies (The Central People's Government of PRC 2007). It included most ministers within the central government and is now chaired by Premier Li Keqiang.[1] A parallel leading group, the National Leading Group on Energy Saving and Emission Reduction, consisted of the same set of members. Although having two different names, the two leading groups were actually one and the same organisation. It could be referred by either name depending on the tasks it carried out (i.e., international negotiation or domestic energy saving policy) (The Central People's Government of PRC 2007). These leading groups had a similar function to that of the cabinet committees in the Australian context, where a group of ministers could make collective decisions (Williams 2014, p. 7). Climate change is a complex issue, and this approach enhanced discussions and debates among relevant ministries which was a positive attempt towards a whole-of-government (WG) approach. Although the possibility of conflict and competition could not be completely eliminated, the collective decision-making approach, in theory, could lead to more rounded and sensible policy decisions (Shirk 1993, p. 7).

China's climate change policies had always been led by the National Development and Reform Commission (NDRC) up until the 2018 administrative reform, when Department of Climate Change (DCC) was relocated under the new Ministry of Ecology and Environment (MEE, rename of former Ministry of Environmental Protection – MEP) (MEE of PRC 2019; MEE of PRC 2019a). This was to say, during the 11th and 12th Five-Year Plan Period (2006–2015), that NDRC played a predominant role in China's climate change affairs. This arrangement was backed up by the argument that climate change was not only an environmental issue but also an economic and developmental issue. NDRC managed China's macro-economy and made important decisions on energy policies; therefore, it had the capacity for better coordination of energy and climate change (Lewis 2008, p. 158). Following NDRC, there were other central ministries which influenced or sought to influence China's climate policies. Among those, the Ministry of Foreign Affairs (MFA), Ministry of Science and Technology (MST), MEP, and Ministry of Finance (MF) were of great significance. They represented the international aspect, the technological aspect, the environmental aspect, and the mitigation methodology aspect of the development of China's climate policies, respectively (Heggelund 2007, pp. 173–175). MFA played an important role in China's international negotiations on climate change.

It had a cooperative relationship with NDRC in the development of negotiation strategies. MST worked closely with NDRC providing technological support for NDRC's climate-related policies such as renewable energy and clean cars (Chang et al. 2003, pp. 463–465; Heggelund 2007, p. 174).

Contrarily, MEP and MF were the two ministries that held opposite opinions towards NDRC's absolute authority on the management of climate change during this period. MEP did so due to a range of practical obstacles caused by NDRC's poor economic planning when fulfilling its environmental protection duties. As China's atmospheric pollution worsened, MEP began to bear increasing blame from the public. However, MEP, as a weak environmental ministry, had little power in the planning and the approval of polluting enterprises. The majority of its work was described as "end of pipe", for instance collecting fines after the pollutants were emitted into the atmosphere. NDRC took absolute charge of China's economic planning and climate change affairs, however, MEP regards some of NDRC's economic planning as contradictory to the principle of climate change mitigation and air quality preservation (Interviewee 3, 2013). As a result, after gaining full-ministry ranking in 2008, MEP started questioning some of NDRC's controversial approvals of heavy polluting projects by taking a firm stand on Environmental Impact Assessment, and openly demanding a voice in the planning process of industrial projects which had high impact on air quality.

The relationship between the MF and NDRC on the issue of climate change could be described as competition for administrative power. MF argued strongly for a carbon tax. It disagreed with NDRC's trial ETS even after carbon trading pilot program was implemented. The reason why MF insisted on a carbon tax but was against carbon trading was that taxes were managed by MF. The management of carbon tax can broaden MF's administrative power and portfolio (Williams 2014, p. 8). So here, special attention is paid to the analysis of the conflicts between MEP and NDRC, and competition between MF and NDRC to understand the dynamics of China's government institutions in the climate-related policy domain during the 11th and 12th Five-Year Plans.

The status of NDRC

The NDRC was the most significant top-ranking ministry responsible for China's macro regulation. Its major responsibilities included making long-term development strategies and important socioeconomic policies, improving economic structure, approving significant national projects, and supervising overall economic reform. NDRC also took charge of a wide range of specific social and economic affairs which included the research, development, and supervision of the implementation of China's climate change policies before 2018 (NDRC 2018, n.p.). Although NDRC is still currently the most powerful ministry within the State Council and oversees a wide range of socioeconomic affairs, the ministry's authority showed an overall declining trend (Song et al. 2008, p. 56). According to the Director of NDRC, Xu Shaoshi, in 2013 alone, NDRC delegated 416 types of approval authority to lower levels of government. The purpose of this was for NDRC to better focus on the most important tasks such as the supervision of the macro-economy.

The predecessor of the NDRC was the State Planning Commission, established in 1952 (Yang & Xia 2014, n.p.). As the name indicated, it was set up to manage China's planned economy and controlled almost all aspects of socioeconomic affairs from national goals at the top, such as setting up annual national GDP targets, to the most trivial details of people's lives, such as determining the prices of basic consumable goods (Worthley & Tsao 1999, p. 573). The functions of the State Planning Commission gradually changed after 1978 when China decided to switch its planned economy to a socialist market economy. During the 1998 administrative reform, the State Planning Commission was renamed as the State Development Planning Commission. The new commission's main duty focused on managing the national economy and exploring developmental strategies. Its intervention into the micro economy was reduced (Ngok & Zhu 2007, p. 227). In 2003, after Premier Wen Jiabao took over, the State Development Planning Commission was renamed again as the National Development and Reform Commission. Since then, the word "planning" was completely abolished from the Commission (Song 2009, n.p.). In 2008, the NDRC gave up its authority over relevant industrial management functions and the State Tobacco Monopoly Bureau. Meanwhile, the NDRC started supervision of the newly established National Energy Administration (State Council of PRC 2008; Downs 2008, pp. 42–43).

When NDRC was responsible for both energy and climate change governance (2008–2018), overlapping responsibilities and fragmented authority existed within the commission (Huang 2014, n.p.). Shapiro (2012, p. 59) identified fragmentation as a common feature of Chinese bureaucracies – both horizontally and vertically. This phenomenon was demonstrated in the data collected from Interviewee 1 at the NDRC. According to Interviewee 1, the disagreement over how to manage climate change mitigation and energy conservation caused a series of coordination problems within the ministry and between central and provincial NDRCs. He highlighted the conflicts between three departments within the NDRC:

> Both climate change policies and energy policies are led by NDRC. There are three major departments involved, the National Energy Administration (NEA), the Department of Resource Conservation and Environmental Protection (DRCEP) and the Department of Climate Change (DCC). These three departments have different preferred strategies and methods in managing climate-related energy policies, as a result, the standard cannot be unified. Some are in favour of calculating energy savings, some are in line with the control of total energy consumption;

> while others may be leaning towards the control of the total amount of CO_2 emission or its intensity. The coordination among these three departments is far from harmonious as all of these departments stick to their own preferred criteria in their policy measures. Moreover, the multiple criteria makes it difficult for lower levels of government to follow, as they need to undertake similar but not completely matched tasks and report to different departments.

The overlapping tasks among these three departments within NDRC undoubtedly increased the workload for the provincial and local governments. The poor coordination was caused by a combination of factors. First, the DCC was a relatively new department. It was not established until 2008 (Low Carbon of China 2011, n.p.). Before the DCC was established, the other two departments had already governed in the fields related to climate change such as renewable energy and energy resource conservation. Therefore, it was inevitable that some tasks of DCC overlapped with the other two existing departments. Second, dealing with climate change involved the management of CO_2 and other minor greenhouse gases. However, CO_2 was generated by almost all social and economic activities. As a result, it was unavoidable for the DCC to interfere with the other departments, especially those with big mitigation potential such as energy and resources. Therefore, solving the overlapping problem will require a WG approach. Finally and most importantly, the overlapping of responsibilities and fragmentation have been a persistent ailment within the Chinese government and require more fundamental administrative reform to overcome.

Back in 2005, the General Office of the State Council announced the System of Law Enforcement on Administrative Responsibility, which aimed at clarifying responsibilities and avoiding negative competition among government departments. The State Council also identified a new implementation method of "disclosing power lists mechanism", in which the responsibilities and authorities of all government departments of different levels were listed clearly on paper. A department could not exercise authority over affairs that were not listed in its own responsibility list (General Office of the State Council 2005, n.p.). However, due to the complexity in the arrangement of the existing administrative system, a completely clear and defined responsibility system proved challenging in reality (Huang 2014, n.p.).

As a matter of fact, the "disclosing power lists mechanism", in the case of climate change governance, had imposed a negative effect on the coordination of the three climate-related departments within NDRC. The clearly defined, similar but slightly varied tasks for the three departments were not only a waste of administrative resources but also increased the workload of lower levels of government and relevant industries. If the three departments could coordinate with each other and work out unified criteria and measures for climate-related energy policies, their own efficiency and the efficiency of their provincial and municipal offices could, in theory, be improved. In 2018,

DCC was moved under the supervision of MEE. A new ministry – Ministry of Natural Resources – was also established. Would this new situation make the coordination among energy, climate change, and environmental governance more complicated? This will be discussed in detail in Chapter 6, as part of the policy and actor updates during the 13th Five-Year Plan.

Status of the MEP

As stated previously, China is unique in terms of its government structure and political system, and the power distribution within the government is extremely complex, fragmented, and uneven (Shapiro 2012, p. 59). Lieberthal (2004, p. 186) describes China's multilevel bureaucratic system as a "Matrix Muddle", which vividly reflects the power fragmentation between central, provincial, and local governments and among different departments at each level. In the centre, different ministries have different rankings. For instance, NDRC is a full cabinet-ranking ministry with top tier authority. There are also ministries with lower rankings which indicate their less significant authority. The ministries with higher rankings in the centre guarantee their provincial and local branches a higher ranking over those with lower ranking parents in the centre. Therefore, government offices with lower rankings do not have authority over those with higher ranking and those of the same rank cannot manage each other (Ma & Ortolano 2000, p. 33).

China's environmental protection sector, compared with the planning and economic focused ministries, had been in a lower position in the central government for many decades. This lower position not only determined the lower ranking of its sub-branches within provincial and local governments, but also prevented it from being involved in major policy development in the centre (Jahiel 1998, p. 758). However, this situation has changed gradually over the past two decades. Contrary to the declining power of the NDRC, the overall trend of China's environmental protection sector had been rising, except for occasional setbacks. Since China attended the 1972 United Nations Conference on the Human and Environment in Stockholm, the rank of China's environmental sector had raised six times (Dwivedi & Vajpeyi 1995, p. 74). This signalled that China had gradually realised the importance of environmental protection.

In 1974, the National Environmental Protection Office was set up. The authority of an office was much lower than that of a full cabinet ranking ministry, and it had no right to issue orders to lower levels of government. After the promulgation of China's first environmental protection law in 1979, local environmental protection bureaus (EPBs) were established. They had direct access to local government leaders and the power to organise meetings and the right to establish specific sub-units. The status of China's environmental sector slightly improved as a result. However, during the 1982–1983 bureaucratic restructuring, which aimed to reduce the size and expenditure of the government, the Environmental Protection Office in the central government was positioned under the newly created Ministry of Urban and Rural Construction and Environmental Protection. Such an arrangement seriously restricted the independence of the administration of China's environmental protection sector and created obvious conflicts between development and environmental protection within the ministry (Jahiel 1998, p. 768).

The consequence of the 1982 restructuring was seriously questioned by environmental personnel and scholars during the Second National Environmental Protection Conference 1983–1984, where the deteriorating status of China's environment raised great concerns. In dealing with this issue, the government set up an inter-organisational body in 1984, the Environmental Protection Commission, which specialised in facilitating communications between EPB and industrial ministries. This improved the coordination between the environmental protection sector and the industrial ministries to some extent and gave relative independence to the environmental protection sector. In the same year, the Environmental Protection Office was upgraded to the National Environmental Protection Bureau with a higher rank than other second-tier organisations (Jahiel 1998, p. 769). The concept of environmental protection was at this time set as a basic national policy and a national regulatory framework was established (Mol & Carter 2006, p. 152).

The National Environmental Protection Bureau gained complete independence from the Ministry of Urban and Rural Construction and Environmental Protection in 1988 and was renamed the National Environmental Protection Agency (Mol & Carter 2006, p. 152). But unfortunately, during the 1993–1994 government administrative reform, the local and county level EPBs were largely cut for the purpose of reducing government expenditure, resulting in heavy polluting Township and Village Enterprises (TVEs) at local level operating unchecked (Shi & Zhang 2006, p. 274).

It was not until March 2008 that MEP gained full cabinet ranking. This rearrangement was partly due to the pressure from the preparation for the 2008 Beijing Olympic Games, where a sound environment would present a positive first impression of China to foreign visitors. But more importantly, China's environmental situation had reached a tipping point where water, air, and solid waste pollution were almost irreversible (Huang 2008, n.p.). This posed serious threats to public health and long-term environmental and economic sustainability. As former Vice Director of the State Environmental Protection Agency (SEPA) Wang Yuqing (cited in Li et al. 2007, p. 39) said:

> The worsening ecological environment has brought about huge economic losses. It also threatens human health and affects social stability. If the traditional practice continues, the environmental situation will continue to get worse.

Facing the worsening environmental deterioration, the former government leaders (President Hu Jintao and Premier Wen Jiabao) identified environmental protection as an important component of the proposed Harmonious Society Concept.[2] In order to adjust the power imbalance between China's economic development and environmental protection and improve China's environmental quality, the Chinese leaders of this era started to devote more effort to environmental protection. Premier Wen Jiabao emphasised that China had to achieve the following three transitions to achieve a sound environmental outcome during the Sixth National Conference on Environmental Protection:

1 The transition from a focus on economic growth to a focus on environment and economic development.
2 The transition from environment as a lagging objective to equal importance with economic development.
3 The transition from the primary use of administrative methods of the [sic] environmental management to a more comprehensive system combining many approaches (as cited in Lan, Simonis & Dudek 2006, p. 8).

Regardless of whether this was sincere or merely rhetoric, it was truly the first time that one of China's top leaders put environmental protection in a position as important as economic development. Therefore, it was of significance for China's environment. However, the pursuit of "equally important" environmental protection and economic development simultaneously was challenging. In reality, China's city smog and climate change issues were largely caused by fossil fuel-powered economic development. The conflict between these two contradictory goals was also reflected at the institutional level in the conflicts between MEP and NDRC.

Conflicts between MEP and NDRC

In March 2008, the MEP gained full cabinet rank and became level with NDRC. However, the same ranking of the two ministries did not automatically endow them with equal power. The challenges that MEP faced, unlike NDRC's internal conflicts, were the lack of power regardless of its rising ranks. In reality, MEP's power was restrained by NDRC on many occasions. According to Interviewee 2, MEP lacked power in important planning and policy decision processes when there were multiple ministries involved. According to Interviewee 3, MEP was only allowed to deal with the "end of pipe" problems and lacked power to do anything during the planning process of polluting industries. Many EPB staff also expressed their frustration about the difficulties they faced in their daily work of environmental law enforcement on public online forums. One EPB officer explained why it was so difficult to collect evidence for enterprises' illegal pollution on the forum *Zhihu*. According to this person, many enterprises' polluting behaviours were random. For instance, some of them truly did emit pollutants illegally when the public reported the cases. However, when the EPB investigators arrived for inspection, they had stopped doing so. This caused great difficulty for EPB investigators' investigation and assessment. Therefore, the EPB officer on *Zhihu* encouraged the public to provide photo and video evidence when they reported to EPBs (Zhihu 2015). For large-scale pollution cases, MEP faced even greater challenges especially when they were backed up with NDRC approvals.

The dispute upon the approval of a 60 billion yuan (approximately US$10 billion) paraxylene (PX) plant at Gulei Village, Zhangzhou City, Fujian Province of southern China, proposed by Dragon Aromatics, provided one such example. On 6 April 2015, 14 people were injured in an explosion at the PX plant. Villagers and houses were impacted for miles. Similar explosions had happened before on the site. The first explosion could be traced back to 30 July 2013. Interestingly, it was rarely known that such a high-impact facility had never completed an environmental impact assessment. Serious conflicts had happened between the MEP and NDRC behind the scenes of the Gulei petrochemical facilities approval process (Wang, He & Wang 2015, n.p.).

Back in October 2012, the MEP expressed serious opposition towards the PX plant at Gulei as it did not pass an environmental impact assessment. In January 2013, MEP fined the plant 200,000 yuan (approximately US$33,000) for starting construction before gaining the relevant environmental approvals. In May 2014, MEP wrote an official letter to urge NDRC to rescind the approval after receiving NDRC's *Approval for Fujian Zhangzhou Gulei Petrochemical Plant Overall Development Plan*. MEP claimed that NDRC's approval process had already violated the *Environmental Impact Assessment Law*. To be specific, NDRC's approval violated Article 12 and Article 14 of the *Environmental Impact Assessment Law*, and Article 16 and Article 22 of the *Planning Environmental Impact Assessment Ordinance*. This official letter was published on MEP's official website, and it was probably the first time that MEP publicly urged NDRC to withdraw an approval (Liang 2015, n.p.).

According to the *Environmental Impact Assessment Law*, "Governments of district municipal level and above or relevant departments of provincial governments should use environmental impact assessment report and its recommendations as important evidence for decision-making in approving specific planning proposals" (People's Congress of PRC 2003, n.p.). The *Planning Environmental Impact Assessment Ordinance* issued by the State Council further indicated that "environmental impact report should be included for consideration for approval; for proposals without environmental impact assessment reports, relevant planning and approval departments should not issue approvals" (General Office of the State Council 2009, n.p.). However, Feng Hai (pseudonym), an expert close to NDRC, stated that NDRC held a different opinion upon the approval for Gulei Petrochemical Plant. In March 2011, NDRC's approval for *Economic Development Plan Zoning Western Side of the Straits* (the so-called big plan) had clearly stated that the planning boundary included the entire territory of Fujian Province. Zhangzhou Gulei

Petrochemical Plant was located within this territory. In March 2014, NDRC approved *The Overall Development Plan of Fujian Zhangzhou Gulei Petrochemical Plant* (the so-called "small plan"). NDRC believed that an environmental impact assessment had already been done for the big plan, and therefore, it was not necessary to have it done again for the small plan (Liang 2015, n.p.).

Although MEP held no authority over the approval of the above project plan, it insisted that it would dismiss any environmental impact-related documentation from Gulei, which were preconditions for approvals according to the *Environmental Impact Assessment Law*. In fact, up till now, MEP has not dealt with any new environmental impact assessment documentation from Gulei Petrochemical Plant. The argument between MEP and NDRC over the approval for Gulei Petrochemical Plant had been going on for a considerable period of time. In the end, NDRC claimed that in order to promote the unity of the two ministries, internal negotiation was the best way to deal with this issue, and the Gulei Petrochemical Plant case was eventually handled quietly (Liang 2015, n.p.). According to the investigation report of the "4.6" Gulei explosion released on 16 August 2015, the CEO of Dragon Aromatics and 12 relevant personnel in charge of production and safety were transferred to the judiciary for investigation. Eleven government officials including the Executive Vice Mayor of Zhangzhou City, Liang Xinwei, Director of Gulei Economic Development Zone, Shen Yongxiang, Deputy Director of Zhangzhou Quality Supervision Bureau, Lu Weilin, and a few health and safety investigators within relevant departments, were punished according to the Party discipline guidelines. Nine managers and responsible surveyors of Dragon Aromatics were removed from their positions or demoted. Meanwhile, Dragon Aromatics was required to improve its health and safety standards. The government of Zhangzhou City was required to clarify responsibility, improve monitoring, and toughen criteria for approvals in order to avoid similar incidents (Shen 2015, n.p.).

The above case showed that the Chinese government's pro-development tendency was unchanged and the NDRC still held the absolute power on economic planning. This status could be confirmed by the following elements of the case. First of all, the personnel who approved the project within NDRC were neither punished nor even mentioned. The officials being punished were those in relatively less significant positions, and the "punishment according to the Party discipline guidelines" is a rather vague outcome. Second, Dragon Aromatics was not ordered to shut down but only experienced minor personnel change, which indicated that NDRC would hold its original approval decision. Finally, toughening criterion for the approval of major petrochemical projects was only mentioned as one of the suggestions rather than the most significant lesson learned. The role and position of MEP were not mentioned at all in the suggestions.

The fundamental reason behind the Gulei Petrochemical Plant accident was that environmental impact assessment had never been truly mandatory for planning and approving departments. MEP always lost the battle when attempting to challenge the authority of NDRC. Although sharing equal administrative ranking, MEP lacked actual power compared with NDRC. As Interviewee 2 noted, "MEP is quite weak in China. As you can see in many government policy documents, MEP is often listed at the bottom among all the involved ministries. This is because MEP does not have authority for approving and decision-making on many important planning cases, only has the right to give recommendations".

This challenging situation for MEP was also reflected through the handling of the increasingly serious issues of climate change and atmospheric pollution in China before the 2018 administrative reform. The interviews that I conducted with MEP and NDRC officials highlighted this point. When being asked whether it was contradictory for NDRC to manage climate change affairs, Interviewee 3 from MEP strongly agreed that it was contradictory and illogical for NDRC as an economic-oriented ministry to manage climate change:

> We need to have a positive and proactive attitude towards the international agreement, so we appointed this task to the most important ministry, NDRC. However, I think developing the economy is NDRC's first task and climate change mitigation is only a by-product. A bigger problem is that there is no constraint for such a powerful ministry. This leads to a situation in which NDRC can do whatever it decides to. If there are two ministries that can constrain each other's behaviour, the second department can argue under the current situation what is the more important thing to do, it will result in more smooth and sensible policies. But unfortunately for the moment, either to develop the economy or to reduce emissions, the priority is totally decided within the NDRC, which leads to great fluctuations of policy.
>
> In many developed countries, environment-related government departments have comprehensive functions. For example, they may have environment, energy and natural resources as one department. Not like in China, our ministry is solely environmental protection. MEP does not intervene nor has the right to intervene at the initial planning stage of a project, only being asked to check its emissions after all pollution has been made. It is totally pointless and ineffective.

Interviewee 1 from NDRC answered the same question with a dramatically different view, believing that only NDRC had the capability to manage China's climate change:

> Energy and energy conservation have always been managed by NDRC. Of course long ago they were managed by the Economic and Trade Commission. The climate change issue has a very close connection with energy consumption and energy conservation, so if you put it under the Ministry of Environmental Protection it will be very difficult to

> coordinate. Moreover, the climate change issue is more of an economic issue. It involves all aspects of the economy, not just pollution. Now we realise that the "end of pipe" pollution control that MEP does is not effective towards carbon emission reduction. The planning of carbon emission reduction must focus on the structure and developmental status of the entire economy. So, I personally believe that putting climate change affairs under NDRC instead of MEP is very suitable. For China, the biggest issue we are facing is the transformation of the economy. If this is achieved, then carbon emission reduction will be achieved automatically. They are of the same direction. Therefore, NDRC managing both economic transformation and climate change makes good sense, at least for the current period of time.

From the above responses it can be seen that both ministries doubt each other's competence in managing carbon emission reductions at the central government level. MEP questions NDRC's ability in making sensible and truly environment-focused policies, while NDRC to a certain extent dismissed MEP's potential in managing carbon emissions at the time of interview. Since 2018, DCC was moved from NDRC to MEE (previous MEP). Will this change give MEE sufficient power to influence on NDRC's decision-making? Or, will it further complex the coordination between energy development and climate change mitigation? These questions will require time to answer. But the reality before 2018 was that the tension between NDRC and MEP branching down to provincial governments further intensified the conflicts between economic development and environmental protection. MEP often fell powerless when issuing orders to provincial and local officials who manage the economy and local EPBs.

Conflicts among MEP, local EPBs, and provincial and local economic departments

NDRC was not the only actor that restrains MEP's administrative effectiveness. During the implementation of many environmental policies, some provincial and local officials who managed economic affairs (including but not limited to provincial NDRC officials) posed even bigger obstacles due to their unwillingness to cooperate with MEP. MEP did not carry out direct and complete management of its provincial and local branches, the EPBs, which was to say EPBs report to both MEP and the provincial governments in their own locations (Jahiel 1998; Mol & Carter 2006; Shi & Zhang 2006;). However, due to the fact that EPBs' funding was provided by the provincial or local government and EPB personnel selection was also influenced by the provincial or local government, EPB officers were often more loyal to the provincial or local government than to their parent ministry in the central government (Jahiel 1998; Zhang 2007). Driven by tax revenue and possible future promotion – one of the most important criteria used for judging

Chinese official's performance was economic growth and significant projects in his or her governing area – the majority of the provincial governments gave priority to economic development. Most project proposals were required to pass environmental impact assessment conducted by MEP before they were handed to NDRC for approval; however, due to the EPB's reliance on the provincial government, many EPB personnel were often forced to disobey relevant environmental laws and gave green lights to proposals that did not meet required environmental standards (Lan, Simonis & Dudek 2006). There was not much MEP and EPBs could do if the planning and approval bodies decided to completely ignore their recommendations, or local governments ostensibly accepted the recommendations but did not carry them out in reality (Liang 2015, n.p.).

Tax revenue from large projects was a great driver for provincial and local officials to choose economic development as a priority. Traced to the root, the Gulei PX plant accident was a tragedy caused by such a motivation. According to the director of the Administrative Committee of Gulei Economic Development Zone:

> Dragon Aromatics is a company that is willing to spend money. They authorised a foreign consultancy for design and consulting, and we have collected over 10 million yuan tax and fees just on this alone. The fine from MEP and the order from NDRC on adjusting raw materials will have only minor impact on the entire project. (Peng & Fang 2013, n.p.)

This was very true. As shown in *Gulei Monthly Economic Report*, the Dragon Aromatic PX Project had reached overall 99% completion, within which designing and procurement were 100% completed and construction was 98.7% completed. A 200,000 yuan (approximately US$34,000) fine from MEP appeared meaningless compared to the 13.780,000,000 yuan (approximately US$2.3 billion) investment of the PX project at Gulei. But unfortunately, this was already the highest level of punishment allowable in the *Environmental Impact Assessment Law*. The cheap fine was also a reason for companies to ignore the authority of MEP.

In fact, not only was the fine cap in the *Environmental Impact Assessment Law* low, penalties contained in many other environmental laws and regulations were also light between 2006 and 2014, which made it cheaper to violate the law than to obey the law. The *Environmental Protection Law* set pollution fine caps for average polluting activity at 100,000 yuan (approximately US$17,000) which suggested significant regulatory failure, while according to the *Water Pollution Prevention Law*, the fine for big pollution accidents could be as high as 30% of the direct financial loss caused by the accident but could not exceed the cap of 1,000,000 yuan (approximately US$170,000). The low caps on pollution fines led to many big polluting enterprises purposely not operating their pollution control equipment. As former deputy minister of the SEPA Wang Jirong (*People's Daily Online* 2009 n.p.) pointed out, light penalties and caps on pollution fines had severely weakened the deterrent effect of many environmental laws. In Wang Jirong's report, a company installed a set of 200,000,000 yuan wastewater treatment equipment. In theory, the company was capable of limiting its pollution level to the required standard. But the operation cost of the equipment was about 100,000 yuan per day which was similar to the cap of the pollution fine. Due to this situation, the company decided not to use the equipment. This was because as the EPB officers could not come to check every day, it was cheaper to pay fines occasionally than operating the pollution treatment equipment on a daily basis. Atmospheric pollution control faced the same problem. The pollution fine for each kilogram of SO_2 was 0.2 yuan; however, if coal-fired power plants treated SO_2, they would need to spend 1 yuan on each kilogram of SO_2, which was five times as expensive as paying fines. Therefore, in order to effectively enforce environmental laws, it is essential to introduce heavier penalties which make the cost of pollution much higher than that of violating the laws. This challenge will be discussed further in Chapter 6.

The power imbalance between MEP and NDRC may not be resolved in a short period of time. MEP's direct supervision of its provincial and local branches is an essential step for addressing this imbalance. In 2018, pilot programs for MEE to directly supervise its provincial and higher branches were carried out. The meaning and impacts of such initiative will be discussed in Chapter 6. Nonetheless, there had been positive move for toughening MEP's authority from the legal perspective. On 24 April 2014, the People's Congress of the PRC passed the new *Environmental Protection Law*. The new law has been in effect since 1 January 2015 (People's Congress 2014, n.p.). The new *Environmental Protection Law* was regarded as the toughest compared with previous versions of environmental protection laws in China. It abolished the fine caps; instead, pollution fines were calculated on a daily basis. EPBs were also granted the authority to order the closure of companies and detain personnel who violated the law (People's Congress 2014, n.p.). Furthermore, this new law enabled the involvement of civil society and the general public in reporting and raising lawsuits towards polluters. ENGOs at the city level or above had the right to conduct public interest litigation towards polluters, and the Minister of MEP announced the establishment of a "WeChat"[3] Reporting Platform, where the public could easily report polluters' illegal behaviours through social media (Zhang & Zhang 2015, n.p.).

According to Zou Shoumin, the Director of the Environmental Supervision Bureau, MEP, the number of pollution cases they dealt with increased dramatically within the first two months of the implementation of the new *Environmental Protection Law*. There were 26 cases of fines on a daily basis nationally, accumulating 12,389,600 yuan (approximately US$2,000,000) of pollution fines. In 527 cases, violators were ordered to close down polluting entities, production was forced to reduce or stop in 207 cases. Relevant personnel in 147 cases were punished accordingly. The enforcement of environmental law improved dramatically since the introduction of the new law. The removal of fine caps, the newly added administrative authority of EPB investigators, and the allowance for public involvement had all triggered this result (Li 2015, n.p.).

Although China's MEP (now MEE) was weak compared with NDRC and still faces obstacles dealing with some provincial and local authorities, the continuous updates of China's environmental laws was one aspect that could potentially strengthen the capacity of MEP. There are still challenges in the enforcement of the new law, such as lack of awareness of the new law, limited resources of EPB investigation teams, and insufficient coordination between the EPBs, the police, and judicial authorities (Wang 2015, n.p.). However, these challenges cannot diminish the significance of the new law as a legal tool to enhance China's environmental protection. There is hope for China's environmental protection sector to become stronger and to have more say in important policy decisions.

Competition between NDRC and MF

Chinese government departments are not merely the implementation bodies of public policies, they are also independent economic bodies that can obtain certain economic benefits from their administrative behaviour. It has been an implicit phenomenon that central ministries keep a proportion of their revenue within the ministries rather than report and hand in the total amount to the central fiscal revenue. Driven by economic benefits, competition among China's central ministries intensifies, especially among those with overlapping authorities on policies or projects with great potential for economic benefits (Xie 2007, pp. 222–223). The competition between NDRC and MF on climate change mitigation was mainly reflected through the argument on decisions of whether to adopt carbon trading or carbon tax. Both ministries attempted to hold total control of China's climate change mitigation approach due to the potential of gaining broadened administrative authority and economic benefit. NDRC emphasised its authority over climate change governance and proposed carbon trading (cap and trade) as China's major climate change mitigation method. The MF, on the other hand, emphasised its authority over taxation and insisted that carbon tax was a more suitable climate change mitigation tool for China's needs (Liang 2012, n.p.).

MF held a firm stance that carbon tax should be China's primary climate change mitigation strategy. Back in 2009, the Financial Science Institute of MF conducted the special report *The Design of China's Carbon Tax System* (China Economic Weekly 2013, n.p.). At this stage, NDRC was still open for discussion between the two options, or at least did not completely eliminate the possibility of carbon tax. According to Deputy Director of DCC, NDRC, Sun Cuihua, NDRC was negotiating with MF on a major strategy for carbon emission mitigation. If carbon emission credits were initially auctioned to enterprises, then there was no need for a separate carbon tax system as the auction itself would have a similar effect to a tax. However, if carbon emission credits were allocated to enterprises free of charge, then there was the necessity for considering a carbon tax (Liang 2012, n.p.). Three years later, a carbon tax was still not put onto the government's policy agenda. According to the Director of the Financial Science Institute of MF, Jia Kangxiang, this situation was due to the top decision makers not reaching an agreement on a carbon tax (China Economic Weekly 2013, n.p.). However, by then, NDRC was already one step ahead of MF by setting up pilot carbon trading programs. Beijing, Tianjin, Shanghai, Chongqing, Hubei Province, Guangdong Province, and Shenzhen were identified as the first seven carbon trading pilots. Even though the initial carbon emission credits were allocated to enterprises free of charge, a carbon tax was not used as a supplementary mechanism. In Sun Cuihua's words:

> Initially, free credits appeared more popular among pilot cities, companies only need to purchase credits if their free credits are running short or to sell credits if they cannot use up the free credits. (Liang 2012, n.p.)

This again showed NDRC's dominance in the governance of China's climate change affairs, because there was no more mention of a carbon tax even though the initial carbon credits were allocated to industries free of charge. Interviewee 2 further affirmed the status of NDRC by revealing the certainty of carbon trading as China's major climate change mitigation strategy in future:

> According to the current situation, it is almost certain that the establishment of a carbon market is a policy priority. The pilots are all big cities of great significance such as Beijing, Shanghai and Shenzhen. These big cities are like the engines of a train. Once they start moving forward, they will lead other cities onto the track. By then, our future policy layout is basically decided. China's public policies almost always follow this rule. If a pilot program proves feasible and successful, then it will become a national policy.

NDRC's decision and implementation of the carbon trading pilot program had excluded MF's role in CO_2 abatement. MF was not satisfied with this result and continued to express its desire for the implementation of a carbon tax. In 2014, just before the Twelfth People's Congress Meeting (2nd session) and the Twelfth National Committee of the Chinese People's Political Consultative Conference (CPPCC, 2nd session), Jia Shen, Head of the Taxation Department, MF, submitted a proposal on converting traditional pollution fines into environmental protection tax, and adding CO_2 into this taxation category. The proposal also requested environmental protection tax and carbon tax both to be collected by local tax departments. However, at that time CO_2 was managed by NDRC; non-CO_2 pollutants were managed by MEP and pollution fines were also collected by MEP. MF's proposal challenged the interests of both NDRC and MEP.

Although the disagreement between MF and NDRC on whether to choose a carbon tax or carbon trading as China's major carbon emission control strategy was primarily interest driven, it did lead to serious discussions over the suitability of the two options for China's specific situation. According to Interviewee 1, research institutes under MF, NDRC, and MEP had conducted numerous studies on a carbon tax, carbon trading, and comparisons of the two strategies. Top universities and semi-independent research institutes, such as Tsinghua University and the Chinese Academy of Social Sciences (CASS), have also conducted research on these topics. There were a few key differences between the two options.

A carbon tax requires the regulatory authority to directly set up a carbon price – the tax rate; while carbon trading requires the regulatory authority to determine the total emission cap and leave it to the market to yield an emission price (Goulder & Schein 2013, pp. 3–4). Some may think that setting up a carbon tax is simpler and cheaper than establishing cap and trade[4] because it is just another tax, and both the government and corporates are familiar with taxes. However, this is not true. In fact, carbon tax and cap and trade require a similar amount of administrative cost as both require the same level of emission monitoring. Just as Keohane (2009, p. 43) put it, "Regardless of whether emissions are taxed or capped, they must be measured". Therefore, cost has never been the centre of the debate of carbon tax or cap and trade; instead, the decisions are usually more out of political concerns. The competition between NDRC and MF, and the final decision on cap and trade in China also reflected political concerns from the international perspective, as explained below.

A cap and trade system may appear more politically attractive in Western democratic countries because it offers something for all stakeholders. Politicians can combat global climate change without using the word "tax". Industry groups and financial intuitions are able to acquire potential profit-adding opportunities in the new market. Environmentalists are in favour of cap and trade as it has a more definite opportunity for a declining cap on carbon emissions (Avi-Yonah & Uhlmann 2009, p. 6). In China, a single-party authoritarian country, the above concerns seemed irrelevant. Why did China not choose a more straightforward alternative, i.e., a carbon tax? Lo (2016, p. 81) argues that the motives behind China's decision in fact stemmed from international politics rather than domestic concerns. The pursuit for "pricing power (定价权)" and "power of saying (话语权)" on the international carbon market could be viewed as part of China's effort towards the competition for global leadership.

This book argues that the Chinese government's decision on cap and trade system is also a reaction towards international criticism on the amount and growing speed of China's total greenhouse gases emission. As mentioned in previous chapters, the main criticism towards China's involvement in global climate change mitigation was that China did not set a cap on its total emission. Even when China promised sizeable reduction targets on its energy intensity and accomplished these targets, there was still criticism of its continuously growing national emissions (Wang 2014). In response to this criticism and in order to further improve its international image, China selected cap and trade as its national emission reduction strategy over its alternative. This is because, "Under a carbon tax, the price for emitting carbon would be fixed, but the level of emissions reduction would not be. With cap-and-trade, there would be a fixed limit on emissions. But the price of emitting carbon would fluctuate depending on ups and downs in the allowances market" (DiPeso 2009, p. 96). By capping its national emission, China left no more opportunity for international criticism for its climate policy. Just as Interviewee 2 stated:

> We like to be praised on the international stage. We want people to think we've done a lot [to reduce carbon emission]. And in fact, we did do a lot. But we have to make them [the international community] believe we've done it.

In the end of 2017, China's national ETS was implemented. This opened a new era for China's climate change mitigation. Since China had already decided to implement cap and trade on a national scale, any further argument over the advantages of carbon tax and carbon trading becomes meaningless. Instead, the focus should be put on how to optimise China's national cap and trade system and get down to the details. These approaches will be discussed in Chapter 6.

Advisory bodies

In ancient China, government bureaucracies were mainly made up of intellectuals, or scholarly officials. Ferdinand (1991, p. 10) argued it was the intelligentsia that moulded the structure of China's imperial bureaucracy. This tradition of valuing intellectuals' advice for government activities and polices was revived after the late 1970s. Since China's reform and opening up, intellectuals had played an increasingly important role in the public policy domain (Goldman 1999, p. 283). This phenomenon was more profound in the process of China's climate-related policies. Relevant research institutes within NDRC were the major advisory bodies on such policies. They were responsible for conducting primary research, drafting policies, and releasing information (Zhu & Xue 2007, p. 453). However, a range of external organisations and personnel were more often seen in the development process of China's climate-related policies including academics, researchers of external agencies, and foreign experts. Williams (2014, p. 9) attributed this phenomenon to the lack of internal expertise within the government. As a result,

China's climate-related policies were often the collaborative efforts of government internal institutes and a wide range of external experts.

The expert community played a very important role in the formation of China's climate-related policies because climate change was a relatively new domain for the Chinese government, and as such ministers and relevant officials were unfamiliar with climate-related policies. Therefore, they relied heavily on the experts for policy advice. Climate-related policy was also complex as it involved all sectors which utilised fossil fuels and emitted carbon emissions. As a result, the government relied on the experts' expertise for technical data and evidence as sources for drafting detailed policies. Furthermore, climate change is a global problem with an international and political aspect. China's navigation of its climate-related policy would attract attention and response from the international community. Due to the concern of international image and face, the Chinese government needed to obtain the best possible advice not only on the technical aspect but also the political aspect of the policy. Therefore, the advisory bodies in China's climate-related policy were of great significance in both technical and political terms.

Official policy research institutes

The National Advisory Committee on Climate Change was the prominent official government agency in climate change policies. It consisted of the most influential experts in this area and reported directly to NDRC and the State Council before 2018 (Williams 2014, p. 9). There were also other official research institutes that contributed to the development of climate-related policies, such as the Energy Research Institute (ERI) within NDRC, the National Centre for Climate Change Strategy and International Cooperation (NCSC), and the State Council Development Research Centre (DRC) (Wübbeke 2013, p. 716). Although there was plentiful expertise within the official formal advisory bodies, NDRC still experienced difficulties in developing and implementing climate-related policies due to the complexity of the climate change issue and the lack of analytical skills of lower level NDRC branches. Interviewee 1 from NDRC described this situation as follows:

> In policies I participated in developing, a range of experts in Beijing from the Department of Resource Conservation and Environmental Protection, National Centre for Climate Change, and Tsinghua University and major industries helped NDRC do the analysis and compose a draft. This draft was then sent to different departments and provinces [provincial NDRCs] to obtain feedback. But there is a problem here, according to my personal experience, I am not here to depreciate or play down anyone, the research ability in the provinces is not sufficient. For example, NDRC sent a policy draft to a province. The provincial NDRC may only oppose it if it is obviously contradictory to the

> province's situation. Otherwise it will do nothing because it does not have the capacity to conduct the detailed analysis to give feedback to the central NDRC. Another problem may be that provincial NDRCs have a general look at the draft and believe there is no problem, but problems emerge once the policy is implemented. But it is already too late. This is why I say the bottom up strength is not sound and provincial research capacity does need to be strengthened.

The interviews I conducted parallel Williams's (2014) findings that the advisory experts on China's climate-related policies were mainly based in Beijing. The above description from Interviewee 1 on the lack of research capacity at the provincial level further echoed this situation. The lack of professional expertise at the provincial and local level creates challenges for policy development at the centre and this challenge will be investigated further in Chapter 6. In the next section, a range of semi-formal external research institutes that also participated in the policy process will be explored.

External advisory bodies

Semi-formal institutes, such as the official government research agencies, were also mainly based in Beijing. Tsinghua University and the CASS were the most frequently involved external advising organs in Beijing, according to Interviewee 1. Although these organisations had connections with the government and were classified by Zhu and Xue (2007) as "semi-formal" think tanks, experts from these organisations were able to conduct independent research and were more likely to provide unbiased analysis on certain policy options. They were often called upon by relevant government departments to participate in the discussion of climate-related issues and their advice was valued.

Feedback from civilian organisations, such as influential NGOs which work closely with enterprises on climate-related issues, was also considered by the government. For example, Quanliang New Energy Chamber of Commerce is an NGO that provides consultancy services to its members in China's renewable energy industry. Interviewee 6 from this organisation indicated that the government was willing to listen to their feedback and recommendations as they had firsthand information on the performance and challenges of member renewable energy enterprises in certain new policies such as Clean Development Mechanism (CDM). However, the channel of feedback of civilian organisations was quite different from that of official government research institutes or semi-formal research organisations. Instead of attending formal government meetings, the feedback of climate-related NGOs was often delivered through business or personal connections with certain government personnel. This was a unique characteristic of how the Chinese government absorbed external feedback. Although it would be ideal if civilian organisations could have a formal channel to express their concerns and feedback, this informal channel could still be seen as progress (Zhu 2009, p. 341). Foreign research institutes and experts, who held no hostility towards China's climate stance, were also included in the process of some climate projects. Their influence could also be enhanced through personal ties with influential domestic experts, research organisations, or personnel in the government.

All in all, China's expert community played a significant role in climate-related policy development. There was a trend that semi-formal research institutes and civilian organisations performed more actively in China's climate affairs. This was partly due to the complex nature of climate change and the insufficient internal expertise, and partly reflected that China gradually opened up its traditional attitude towards climate change. The expert community could be seen as the next most important actor after the government institutions in the formation of China's climate-related policies. The government drew from their discussions and advice to make a decision. However, the insufficient expert capacity at the provincial and municipal level was a critical obstacle for China's climate-related policies to be well implemented. Opportunities to improve this situation will be explored in Chapter 6.

Enterprises

Enterprises were another major player in shaping China's climate-related policies. However, within this category, the attitudes of different types of enterprises towards emission reduction policies were not unanimous. Large energy SOEs and newly emerged private renewable energy enterprises held dramatically different views on such policies. Large coal-dominated energy SOEs only welcome certain emission mitigation policies if government compensation was high and the overall adoption was profitable (Bergsager & Korppoo 2013, p. 12). Medium- and small-scale renewable energy companies, in contrast, tended to embrace such policies in general. The cause of this situation is explained below.

China's energy sector accounts for approximately 50% of the country's total CO_2 emission. This was largely due to the dominance of coal in the country's energy structure (IEA 2011, p. 594). Within the energy sector, coal-fired power plants generate 97% of total CO_2 emissions. According to the IEA's estimate, 57% of China's electricity generation capacity was controlled by SOEs, which was to say, Chinese energy SOEs held great potential in emission mitigation if they could adopt climate-friendly measures (Baron et al. 2012).

China's coal-fired energy generation was dominated by five large energy SOEs, namely China Huaneng Group, China Power Investment Corporation, China Guodian Corporation, China Datang Corporation, and China Huadian Corporation (Shen et al. 2014, p. 350). There were also small-scale power generators managed by provincial and township governments. Together with some private generators, they took up a relatively small proportion of 29.92% of the total generation capacity (SERC 2011, n.p.). Bergsager and Korppoo (2013) argued that SOEs as a key actor for the mitigation of China's CO_2 emissions deserved primary attention from policy analysts in the field of climate policies. SOEs exerted their political and economic influence upon the formation and implementation of climate-related policies through their formal and informal connections with relevant bureaucracies. They bargained for the best possible support from the government, such as compensation, tax breaks, and free technological updates.

During both the 11th and the 12th Five-Year Plans, the above five large energy SOEs had adopted a range of mitigation measures in resource conservation, emission control, renewable energy development, and diversification of energy sources, and had achieved positive results. For instance, China Power Investment Corporation decreased its standard coal consumption rate by 31 g/kWh, and China Huaneng Group had a decrease of 22 g/kWh (Shen et al. 2014, p. 352). However, Bergsager and Korppoo (2013, p. 12) also identified the tendency that SOEs were generally in favour of mitigation measures that were economically favourable but might resist those that were costly and diverged from their current goals. The inconsistency of SOEs' attitudes towards climate change mitigation measures had delayed many of China's ambitious mitigation policies. For instance, two of China's biggest grid companies, State Grid Corporation of China and China Southern Power Grid Company Limited, once openly opposed connecting renewable energy to grids due to concerns for the stability of their grids. In addition, large SOEs, although gradually embracing government mitigation measures, would lobby against big changes in policy so that they had sufficient time to adapt (Williams 2014, p. 15). As a result, the role of large energy SOEs in China's climate-related policies was often as barrier. Large energy SOEs were an actor which had great influence on the Chinese government's decision-making in climate-related energy policies due to their importance to the economy. The government and the advisory bodies had to consider their feedback and suggestions on the development of climate-related energy policies. Therefore, how to encourage large energy SOEs' enthusiasm for more ambitious climate change mitigation is an important question to address, and this will be discussed in detail in Chapter 6.

Medium- and small-scale private enterprises of renewable energies

China's private investment in the energy sector increased steadily and had become an important accelerator for the development of China's renewable energies. According to *National Energy Administration Document No. 179* (NEA 2012, n.p.), China's private hydropower took up 26% of total national hydropower capacity, and private wind power took up 20% of the national total. So far, private investments had led the research and development of solar thermal, bio-energy, PV solar, solar hot water systems, and batteries for solar energy, and also played an important role in the country's manufacturing of wind turbines. The government encouraged private investors to get involved in the research and development of renewable energies by providing financial support, ensuring a fair market environment and offering a favourable grid price (NEA 2012, n.p.). According to the deputy director of NDRC, Chen Deming, the government saw private ownership companies and SOEs equally in the sector of renewable energies, and subsidies were distributed to all eligible companies without asking the ownership type. Based on the current market share of private investment, the government believed that private investment would become the main force in China's renewable energy industry in future (China IRN 2007).

NDRC issued a range of policies to ensure the competitiveness of renewable energy industries which acted as another layer of guarantee for private investors. For instance, all electricity end users had been charged a renewable energy surcharge since 2006. The surcharges collected were used as a subsidy for renewable energy industries for research and development, power generation, system operation, and maintenance (Zhao et al. 2011, p. 26). The government also identified a cost-sharing mechanism for electricity generated from renewable sources since the 11th Five-Year Plan (2006–2010). The mechanism required that grid companies must provide service to renewable energy generation plants and must purchase electricity generated from renewable sources within their grid coverage; the higher feed-in tariff of renewable energies were to be shared by nationwide end users (Wang 2010, p. 706).

These policies played an important role in reducing costs and ensuring profit of renewable energies. As a result, they largely stimulated private investment in the renewable energy sector. According to the Director of the National Energy Commission[5] Expert Advisory Committee, Zhang Guobao, among all listed Chinese renewable energy companies on the top 500 global renewable energy company list, private companies took up approximately 80%. He explained that this phenomenon was due to the fact that renewable energy was a relatively new field, and private companies were standing on the same platform as SOEs and even foreign companies. Innovation and market, rather than government planning, were the main drivers for these private investors. Therefore, what the government should do is to provide the best possible support for them to flourish (Xing 2014, n.p.). However, to implement such a proposal is another thing. Whether or not private renewable energy enterprises could really stand on the same platform as SOEs and enjoy equal support as renewable energy under large energy SOEs is doubtful in reality. The equality issue will be analysed in detail in Chapter 5.

The Chinese government's determination and policies in promoting renewable energy development were genuine. However, the renewable energy industry did face a range of challenges. Although the renewable energy surcharge had been raised seven times from the 0.002 yuan per kWh in 2006 to the 0.015 yuan per kWh in 2014, there was still an accumulated 14 billion yuan (approximately US$2.3 billion) gap of unpaid renewable energy subsidy (Sun 2014, n.p.). This gap had led to delays in handing out renewable energy subsidies to some companies, which made small-scale private companies especially vulnerable. According to Zhu Gongshan, CEO of GCL Group Holdings Limited, the development of China's PV solar industry still largely relied on government subsidies. But in some regions, subsidies were only handed out once a year, which put great pressure on companies specialising in PV solar energy. Companies needed government subsidies to maintain healthy operations. Without sufficient cash flow, they experienced difficulties in equity refinancing. This in turn pushed up the overall investment cost of PV solar power plants and extended the recovery period of investment (Hao 2015, n.p.). The insufficient and delayed subsidies not only put pressure on renewable energy companies, but also created a triangle debt relationship among the government, the renewable energy producers, and the manufacturers of renewable energy equipment. The government owed money to renewable energy producers; renewable energy producers owed money to their equipment suppliers; and the equipment suppliers owed money to the parts manufacturers. This showed that insufficient subsidy funding and delayed disbursement caused major negative impacts on the health of China's renewable energy industry (Sun 2014, n.p.).

Among all renewable energy enterprises, medium- to small-scale companies faced tougher challenges compared with those under major energy SOEs. Large energy SOEs held more formal or personal connections with the government, and were therefore more likely to be successful in the process of bargaining for the unpaid subsidy (Williams 2014, p. 13). Small- and medium-scale renewable energy companies were less successful in this process. In order to ease financial pressure on medium to small businesses, the government took some actions to make up the gap. For instance, the Renewable Energy Division of NEA sent out an emergency letter titled *Please Provide Renewable Energy Subsidy Gap* to relevant renewable energy companies in April 2014. The letter required companies to provide the data of the government's unpaid subsidy accumulated from 2012 to 2014. The letter aimed at paying back the delayed subsidy, analysing the conditions of the renewable energy industry, and rearranging the future budget for renewable energy subsidies (Xinhua News CN Stock 2015, n.p.). Obviously, how to guarantee sufficient and prompt subsidy for the renewable energy industry was a problem that needed to be urgently addressed, and whether or not the government's action could lead to sustainable subsidies remained to be seen. Should the government continue to rely on a renewable energy surcharge to generate the subsidies? Or should there be diversified sources? If diversified sources are in consideration, what should they be and how should they be obtained? These issues will be addressed in Chapter 6. Overall, the renewable energy sector was weak compared with the traditional coal power sector. Currently, renewable energy enterprises under private ownership faced a range of difficulties. But they had the potential to become a more important actor under China's national ETS. China's national ETS has been implemented for over a year by now. The energy industry was the only industry included in the trading. Renewable energies played an important role in both carbon trading and off-setting. Although it is still early to judge the overall impact of the national ETS on China's renewable energy industry, it is foreseeable that renewable energy will become more competitive and profitable in the scheme. It is hoped that the renewable energy actor will be stronger and have more power in China's future climate-related energy policies.

The media, the ENGOs, and the general public

The combined power of the media, the ENGOs, and the general public have long been regarded as an important social force in shaping environmental policies and influencing environmental transitions in Western democracies (Binder & Neumayer 2005, p. 527; Jänicke 1996, p. 71). It is believed that strengthening environmental institutions within the government alone is not sufficient to generate sound resolutions to many complex environmental problems. Rather, the involvement of the media, the ENGOs, and the general public can often push environmental policies to the right direction and result in more rounded environmental solutions (Weidner 2002, p. 1365). In China, the media, although proactive in reporting incidents of environmental neglect, is largely controlled by the CCP. The emergence of ENGOs was late and their growth was slow (Martens 2006, p. 224). The general public had only limited capacity for action against pollution due to a relatively low level of income and education, and many of them were employees of polluting enterprises and relied on them for income (van Rooij 2010, p. 63). Moreover, a traditional belief still persisted among the public that authority should not be challenged if conditions were tolerable (Jing 2000, p. 216, Lin & Gil 2016, p. 145). As a result, the public tends to only react to extremely serious environmental issues such as large-scale chemical explosions and environmental issues that directly impact on their livelihood, such as local water and toxic waste pollutions (Yang 2010, p. 101).

The climate change issue attracted relatively less attention from the Chinese public; however, it was embedded in another environmental problem, air pollution, which was often perceived to be more important as it directly impacts on the majority of Chinese city dwellers' health and wellbeing (Lin 2009). In China, the majority of cities suffered from serious air pollution, and only 1% of city dwellers were lucky enough to breathe air that was adequate according to EU standards (Shapiro 2013, p. 224). According to Kan, Chen, and Hong (2009, p. 187), sufficient evidence proved that China's outdoor air pollution generated negative impacts on public health as shown in increasing hospital cases of lung and respiratory diseases. The public urged the government to take actions towards air pollution, and their voice was highlighted in the production of "Under the Dome", a documentary on China's air pollution and related health problems. The documentary generated nationwide discussion on the origins of city smog (Zhang 2015, n.p.). As a matter of fact, air pollutants and greenhouse gases could be traced back to the same sources in the Chinese context. For instance, heavy reliance on coal for power generation and exhaust from increasing numbers of passenger cars result in both particulate matters and greenhouse gases. Therefore, in theory, any public actions against air pollution would have the potential to simultaneously push forward China's climate change policies. However, the reality was that impactful actions in fighting against air pollution initiated by the media, the ENGOs, and the public were minimal.

The media worked at the forefront fighting against environmental pollution. Chai Jing's documentary "Under the Dome" on city smog attracted at least 200,000,000 views within the first 48 hours. This was to say, one in three Chinese web users were watching Chai Jing's documentary, and there were 632 million web users in China. In theory, the high viewing rate was a good thing for awakening the public's environmental consciousness and for attracting the government's attention to solve the problem. However, things went in the opposite direction. The documentary was banned at midnight on 2 March 2015, two and a half days after its initial broadcast. The official media outlet Xinhua News deleted the articles on the comments the Deputy of the People's Congress made on "Under the Dome" and the message that the Minister of MEP wrote to Chai Jing. It also urged other media outlets not to use this content. By then, the documentary was no longer accessible on the internet (Wong 2015, n.p.).

The official explanation for the disappearance of the documentary from the internet was that its copyright was sold to Phoenix TV, whose headquarters was based in Hong Kong. Therefore, it could no longer be viewed in mainland China. However, the particular time for the ban directly overlapped with the gathering of the National People's Congress (NPC) and the CPPCC (Two Sessions), which inevitably left people thinking there was more to it. The mainstream topics of the Two Sessions had always been around the positive achievements of China's reform and development. Chai Jing's documentary on air pollution, although filled with love and humanistic care, talked about a "negative" social issue. Therefore, it was not a Two Sessions friendly topic. Moreover, the documentary had already gained a huge viewer base. If people got the chance to dig deeper for the fundamental cause, they would point to China's coal and petroleum industries and the lack of action of MEP, which might potentially cause large-scale social unrest. This was what the Chinese government wished not to see. The prompt banning of "Under the Dome" showed just how much control that the Chinese government had over the media. The Chinese government had a strict limit to its tolerance for media reports on environmental issues. It was acceptable to report on the habitat loss of an endangered monkey, but it was not acceptable to report on air pollution damaging hundreds of millions of people's health. From the above incident, it could be observed that the Chinese media faced tough political restrictions when reporting on the climate-related air pollution issue, because it simply involved too many people, which tested the Chinese government's boundaries.

Actions of many Chinese ENGOs towards air pollution were limited to recording air pollution levels, informing the public about the harm of air pollution, and providing education to citizens (Zhao 2015, n.p.). Naming and shaming top polluters who contributed to city smog could be regarded as the most radical of ENGO's actions (Lei 2014, n.p.). Also, the Chinese public's opposition towards air pollution was mainly expressed verbally through social media but in daily life people tended to tolerate or adapt to air pollution rather than protest against it. This could be reflected through the boom of the so-called "smog economy" – the record high sales of masks, air purifiers, and "escape from smog" tours[6] (Wen 2013, n.p.). There were a number of reasons behind the weak performance of Chinese civil society on the issue of air pollution and climate change, among which the lack of targets for action, the practical challenges of Chinese ENGOs, and fractured public opinion were significant.

Lack of targets for action

The causes of city smog, similar to the causes of climate change, are complex. It may include a mixture of industrial waste gases, emissions of coal-fired power plants, dust from construction sites, car exhaust, and the burning of straw in the surrounding countryside (Molina & Molina 2012, p. 664). According to Sun Junying, the Director of the Institute of Atmospheric Composition, Chinese Academy of Meteorological Sciences, Beijing's fine particulate pollutants (PM2.5) included approximately 40% organics, 17% sulphate, and 14% nitrate. The Deputy Director of the same institute, Wang Yaqiang, indicated that the origin of PM2.5 was complicated. It included both fine particulates directly emitted from coal burning, car exhaust, and fugitive dust, and secondary fine particulate matters transformed from complex chemical reactions of sulphur dioxide, nitrogen oxides, and volatile organic compounds in the air (Zhi 2013, n.p.). Therefore, it is very difficult for ENGOs or the public to identify a particular target to take action against. According to van Rooij (2010, p. 58), citizen actions were determined by two prerequisites: "naming" and "blaming". "Naming" required citizens to understand the seriousness of the pollution, and "blaming" required citizens to attribute the damages of the pollution to a responsible entity. Here, understanding the seriousness of the pollution was not simply the awareness of the harmful impacts, but how well they knew the impacts. Unfortunately, such knowledge was usually gained slowly among citizens if a particular type of pollution did not pose sudden and direct impact on local environment and public health (Alford et al. 2002, p. 498). China's air pollution was a case with such characteristics. The knowledge of air pollutants' origins was complex and the health impact might take a relatively long time to manifest. Both of these characteristics added complexity and difficulty to the process of "naming" and "blaming", and thus, posed barriers for citizen action against air pollution.

Constraints on Chinese ENGOs

Chinese ENGOs faced more constraints and obstacles compared with ENGOs in the West. These included the initial government screening for registration, the annual renewal of registration, the lack of funding and quality staff, the lack of access to pollution data, the difficulty in pursuing lawsuits, and overreliance on the power and publicity of ENGO leaders (Schwartz 2004, pp. 38-39). Therefore, even if Chinese ENGOs or the public had the ability to identify major air pollution sources, they still faced many practical difficulties in challenging big polluters. Among all the challenges, political constraints and difficulties in pursuing lawsuits were the two major ones.

Many of China's large-scale environmental organisations are government-organised non-governmental organisations (GONGOs). GONGOs are established by government agencies. It is common that such GONGOs' funding and senior personnel are from their sponsoring government agency. Chinese GONGOs do not function like grassroots organisations and their audiences are mainly targeted at government officials, policy makers, and foreign NGOs which can provide funding for China (Knup 1997, p. 11). Therefore, it is unrealistic to expect Chinese GONGOs to act the same way as Western ENGOs which are independent from the government. Western style ENGOs do exist in China. Wu (2002, p. 37) defined grassroots type of Chinese ENGOs as "private institutions that do not draw on state funding or utilise state assets, but do engage in non-profit social activities". Such ENGOs tend to be organised by individuals, small in scale, and of greater independence; however, they still have to follow the regulation and accept the supervision of relevant government departments (Knup 1997, p. 12).

Chinese NGOs are legally required to register with the Bureau of NGOs Administration within the Ministry of Civil Affairs. Before the registration, an NGO must undergo a screening process by a government department. The screening result then goes back to the Ministry of Civil Affairs for consideration of approval. If the registration is approved, the NGO will be supervised by the government department which performed its initial screening on a range of political affairs, financial, and daily activities (Otsuka 2002, p. 231). Registration of Chinese NGOs has to be renewed annually and renewal can be rejected if an NGO's behaviour is not in accordance with the *Regulation of NGO Registration and Management*. To be specific, the government has the power to close down an NGO if its activities are interpreted as follows:

1. Opposing the principles of the constitution.
2. Opposing the leadership of the CCP.
3. Harming national unity and security.
4. Harming national interest, social interest, or interest of other organisations and citizens.
5. Carrying out activities that are contrary to social and public morals
6. Developing sub-branch offices (Wu 2003, p. 38).

Due to the wide range of constraints and the government's absolute control of their survival, Chinese ENGOs tended to act carefully without rigidly criticising the government's failures in managing the environment. Many Chinese ENGOs even purposely limited their size due to the concern that the expansion of membership and networks might be considered by the government as contravening government policies (Wu 2003, p. 39). Interviewee 4, a local EPB officer, reflected the difficult situation of Chinese ENGOs in relation to the political constraints they operate within:

> Why don't we have radical ENGOs? You should ask why we should have radical ENGOs. Are they effective in China? Do they do any good? Even if you set up a small green party like in the West, can you butt in the parliament and balance the power of the two major parties? If you can't, then you shouldn't do so. This is understood by the rich and the poor.

The above statement ironically pointed out that China's political system was the fundamental restraint for Chinese ENGOs to develop and act like the Western ENGOs. This political condition was unlikely to change in the foreseeable future, which meant Chinese ENGOs were unlikely to strengthen within a short period of time. The level and scope of operation of Chinese ENGOs were eventually determined by the government.

If political restraint was a "congenital defect" of Chinese ENGOs, then legal barriers would be the hardship that many Chinese ENGOs had to endure. Environmental public interest lawsuits were commonly regarded as the most effective way for ENGOs to take action against polluters (van Rooij 2010, p. 67). Although China's environmental laws and regulations were improved and strengthened over the years, they still contain weak, impractical, and incomplete elements (Alford & Liebman 2000, pp. 740-743). When pursuing lawsuits against polluters, ENGOs faced a range of legal obstacles in the process of opening a case, which included providing evidence of damages, demonstrating the existence of a polluting act, and finding evidence for causation (van Rooiji 2010, pp. 60–69). They also faced practical challenges such as high legal costs (Richerzhagen & Scholz 2008, p. 319). As a result, there were very few cases in which Chinese ENGOs used the legal tool to fight against the polluters. In cases where ENGOs did pursue legal actions, the challenges they faced were enormous.

The first environmental public interest litigation organised by a grassroots organisation, the Yunnan Qujing chromium slag pollution case, was one such example. In June 2011, two contractors of Yunnan Luliang Chemical Industry Co., Ltd. illegally dumped over 5000 tons of chromium slag on the hill slopes and along the roadsides of Qilin District, Qujing City. The chromium slag caused severe toxicity to the Nanpan River and killed 77 cattle that drank from the river (Li & Wang 2012, n.p.). The case was initiated by the Friends of Nature and Chongqing Green Volunteers' Union in October 2011. The members of the Friends of Nature initially encountered violent insults from the security guards of Yunnan Luliang Chemical Industry Co., Ltd. during sample collection. They were also given the cold shoulder by the local government. After this, the court required water and soil samples to be examined by environmental institutions with the qualification of judicial expertise. However, environmental institutions with such expertise were extremely rare in China. In the end, Yunnan Luliang Chemical Industry Co., Ltd. broke the settlement agreement proposed by the Friends of Nature at the last minute. So, the case went into a new round of trial procedure and judicial appraisal after the two-year battle. The Friends of Nature and Chongqing Green Volunteers' Union were not able to afford the cost for new court action. They planned to borrow 1 yuan each from one million citizens, and the money would be repaid by the defendant if they win the case (Li 2014, n.p.). This demonstrated how slow and difficult an ENGO-led environmental lawsuit was.

On 1 January 2015, the new *Environmental Protection Law* replaced the old one and took effect. The implementation of the new law brought some new hope to China's ENGOs since it emphasised the importance of public participation. It also clarified the responsibility of polluting industries and relevant government departments in providing the public with transparent pollution data. Article 57 and Article 58 clearly identified the public's right in reporting polluters to the government and eligible social organisations' rights in initiating lawsuits against polluters at the People's Court (Xinhua News 2014, n.p.). Many ENGOs intended to test out the effectiveness of the new law by pursuing environmental public interest lawsuits against polluting entities. However, there was no case against air pollution by the end of the 12th Five-Year Plan (Li 2015, p. 34). Apart from political and legal obstacles, the Chinese public's fractured opinions towards air pollution and climate change made them unlikely to take united, joint action towards the problem.

Fractured views of the Chinese public

The internet provided an important channel for the public to comment and debate on environmental issues in a range of public forums, and the relative anonymity of the internet encouraged citizens to do so (Martens 2006, p. 212). However, in many cases, the internet does not necessarily function to form a united and positive public opinion or push forward innovative public action against pollution. Take the documentary "Under the Dome", for example. Although it raised public awareness on air pollution, it did not achieve the goal of inspiring citizens to take action towards city smog. Instead, the focal point of air pollution was soon buried under the discussions of the producer and presenter Chai Jing's characteristics, her motivation, techniques in seeking publicity and debates on the truthfulness of scientific data presented in the documentary (Huang 2015, n.p.). This outcome could be explained by a combination of China's current political and social realities.

First of all, China's recent public opinion atmosphere was flooded by the trend of moral criticism. Moral criticism become the mainstream official discourse towards corrupt officials, foreign influences perceived to have negative impacts on China and controversial celebrities and professionals (He, Thornton & Cheng 2015). This official trend in mainstream discourse heavily influenced the public's views on many issues, which, in Chai Jing's case, easily shifted the focus of the discussion from air pollution to the presenter herself. In many of China's influential public forums, such as tianya.cn, baidu tieba, and weibo, there were many threads which commented on Chai Jing's motives and personal qualities rather than the air pollution itself. Some regarded Chai Jing as a traitor because of her migration to the USA and believed she should therefore stay out of China's affairs. Some suspected that Chai Jing's true intention was to get publicity. There were also negative comments about her characteristics and previous personal lifestyle. Some even took pleasure in Chai Jing's misfortunate daughter who was born with cancer, stating that the cancer was not caused by city smog but by her smoker mother's immoral celebrity lifestyle.

Second, the ever-enlarging gap between the rich and the poor resulted in the so-called hate-the-rich trend. This emotion among the poor generated radical comments on the inequality of the living environment and consumer goods between the rich and the poor, which further shifted the focal point of the discussion. There was one type of comments which ironically stated that city smog was good. This was because officials and the rich could buy imported bottled water to get away from water pollution, but they could not buy foreign air and had to breathe the same haze as the poor did.

Third, increasing cases of social injustice stimulated cynicism within the public opinion sphere, which led to the public's distrust on a range of media releases. This was also why many Chinese citizens started to question the true motives behind Chai Jing's documentary soon after the initial wave of discussion of air pollution. Moreover, there were cases where arguments raised by individual intellectuals were seriously and obviously misleading. Some experts' statements even hurt the general public's feelings, such as the theory that city smog was caused by Chinese style cooking and barbecuing (Huang, Ren & Liu 2013, n.p.). This further weakened people's trust in professionals.

The weak role of ENGOs and the public in combating climate change and air pollution was caused by a combination of constraints. The origin of atmospheric pollution is complex. It involved all enterprises and individuals who consumed energy and produced emission. However, the major contributors to China's serious atmospheric pollution were the carbon-intensive energy structure and the inefficient production model. The political constraint was something that was impossible for Chinese ENGOs to overcome in the current political situation. However, there was hope for improvements in new environmental laws, which would better facilitate ENGOs to fight against pollution through legal action. The fractured public opinion on the problem of atmospheric pollutions was another obstacle to achieve organised action towards pollution. To overcome this problem, education alone was not enough. It required the government, business, intellectuals, and influential professionals to reestablish positive moral values, show true consideration for the people, and solve real problems for them. However, it has to be realised that the constraints that Chinese ENGOs and the general public faced were tough and would take time to resolve, and almost all of them would require the government to lead the change. This explained why civil society played a much less important role than the government in solving China's atmospheric pollution problem. Even the Chinese ENGOs were well aware of this situation. Just as Interviewee 5, an ENGO officer in Beijing, stated:

> The extent that the public and ENGOs can participate in treating city smog is actually quite limited. It has to be a top-down process using administrative tools. It requires government monitoring and innovation in pollution controls. What can we teach the public? Drive less? Eat less barbecue? Or set off fewer firecrackers on Chinese New Year? Working in the environmental field myself, I've seen many civil environmental actions, especially those organised by university or high school students, which are basically fantasising and hardly have any impact on the environment. So the more I see, the more I believe that city smog has to be solved with tough administrative measures.

What the roles of the above actors mean

Among all the actors this chapter identified and analysed, the Chinese government played the most dominant role in the development of China's climate-related policies. This is to say, in order to improve China's climate-related energy policies, the government actor has to be strengthened. China's climate-related energy policies followed a top-down approach. It required the government to take the lead and to work in a more efficient and coordinated way to develop more sensible, effective, and rational policies. During the 11th and 12th Five-Year Plan period, the relevant departments in the climate policy domain were fragmented with overlapping and competing responsibilities.

Among all ministries that were involved in or wished to be involved in China's climate-related policy decisions, NDRC was the ultimate authority. However, it was unwise to exclude MEP from the climate policy process as China's air quality worsened and the public's voice soared on treating air pollution. Since 2018, MEE took over the management of climate change from NDRC. It is anticipated that the leadership of MEE on China's climate change mitigation truly values integrity and quality and contributes to sensible and workable policies.

In terms of the debate between MF and NDRC continuing over whether to choose carbon tax or carbon trading as China's primary mitigation strategy, although was initially beneficial for generating discussions and investigating the strengths and weaknesses of both mechanisms, the discussion should not be overly lengthy. The government should realise that it should give out a clear signal to enterprises on the definite policy decision of either a carbon tax or carbon trading or the two combined. Overly lengthy debate will cause distraction and confusion among companies that have participated or continue to be participating in the national ETS.

The coordination between the central government and provincial and local government was also not ideal. Better coordination was needed among horizontal departments. The government should also strengthen climate expertise in provincial and municipal governments. This was because the expert community, who acted as the adversary bodies for the government, played the second most important role. They were the main force in the theory, debate, and drafting of climate-related policies. Their capacity and involvement in the provincial and municipal level determined how the central government's policy was understood and implemented. Currently, such capacity was insufficient at the provincial and municipal level. Therefore, this needs to be urgently addressed.

The energy industry, especially large energy SOEs, was a powerful actor which had the ability to shape China's climate-related energy policies. Their bargaining power and connections with the government meant their interests and the extent of willingness to mitigate had a strong influence on the final policy decision. Therefore, to encourage large energy SOEs to adopt more ambitious goals and at the same time restraining their overly powerful bargaining power would have a positive effect on the outcome of China's climate-related energy policies. The renewable energy industry held the key to China's long-term energy structure optimisation and climate change mitigation. Therefore, it was an actor with significant future perspective in China's climate-related energy policies. The healthy development and increasing strength of this actor will have a positive impact on China's future climate-related energy policies, especially under the implementation of the national ETS. The media, the ENGOs, and the public played only a limited role due to a range of political, legal, and practical obstacles. However, they did fulfil their responsibility in raising public awareness of air pollution and climate change. In order for the Chinese civil society to function better within the current political limits, the media and ENGOs should push forward for simpler and clearer procedures for civil environmental lawsuits.

Summary

This chapter identified the main actors in China's climate-related energy policies, namely the relevant government departments, the expert communities, the energy industries, and the civil society. It analysed the power and influence of each actor on China's climate policy. It was important to understand the strengths and weaknesses of these actors because they had different degrees of influence on policy outcomes. This chapter painted a picture of competing interests and agendas, and fragmentation in China's climate-related policy development. The following chapters demonstrate how these themes played out in the context of specific policies, namely the CDM and Energy Conservation and Emission Reduction Scheme (ECERS).

List of Interviewees

Interviewee 1, Officer of NDRC

Interviewee 2, Research fellow at Research Centre for Sustainable Development, CASS

Interviewee 3, Officer of MEP

Interviewee 4, Local EPB officer

Interviewee 5, ENGO officer in Beijing (Organisation name wished not to be identified)

Interviewee 6, ENGO officer Quanlian New Energy Chamber of Commerce

Notes

1. Members of the Leading Group: Zhang Ping (Deputy Secretary of the State Council), Ma Kai (Director of NDRC, Director of the Energy Office), Yang Jiechi (Minister of Foreign Affairs), Wan Gang (Minister of Science and Technology), Zhang Yunchuan (Director of the National Defense Science and Technology Committee), Huang Shuxian (Deputy Minister of the Ministry of Supervision), Jin Renqing (Minister of Finance), Xu Shaoshi (Minister of Land and Resources), Wang Guangdao (Minister of Construction), Liu Zhijun (Minister of Railways), Li Shenglin (Minister of Transport), Chen Lei (Minister of Water Resources), Sun Caizheng (Minister of Agriculture), Bo Xilai (Minister of Commerce), Gao Qiang (Minister of Health), Li Rongrong (Director of the State-owned Assets Supervision & Administration Commission), Xie Xuren (Director of the General Administration of Taxation), Li Changjiang (Director of Bureau of Quality Control), Zhou Shengxian (Director of Environmental Protection Bureau), Yang Yuanyuan (Director of Civil Aviation Administration), Xie Fuzhan (Director of Bureau of Statistics), Jiang Zhibang (Director of Forestry Bureau), Jiao Huancheng (Deputy Secretary of the State Council and Director of National Government Offices Administration), Cao Kangtai (Director of Legal Affairs Office), Bai Chunli (Vice president of Chinese Academy of Sciences), Zheng Guoguang (Director of Bureau of Meteorology), You Quan (Chairman of State Electricity Regulatory Commission), Sun Zhihui (Director of State Oceanic Administration), Jie Zhenhua (Deputy Director of NDRC).
2. Each generation of Chinese government top leader would produce an ideological work, for example Mao Zedong Thought, Deng Xiaoping Theory, Jiang Zeming's Three Represents, and Hu Jintao's Harmonious Society.
3. WeChat – the most widely used social networking app. in China with over 600 million users worldwide.
4. Cap and trade programs designed to regulate air pollution emissions share the same basic structure. The government establishes a cap on the total emissions of a certain pollutant from a set of regulated sources over a fixed compliance period. The cap is then divided into allowances, each representing the right to emit a unit of the pollutant. The government allocates allowances among the regulated sources, and the

sources may then trade them. At the end of the compliance period, each regulated source must show that it holds a sufficient number of allowances to cover the units of pollution that it emitted (McAllister 2009, p. 398).

5. The National Energy Commission was established in 2010 as a higher level coordinating committee for the National Energy Administration. It is currently chaired by Premier Li Keqiang and Deputy Premier Zhang Guoli. The 20 committee members are Deputy Secretary of the State Council, Minister of Foreign Affairs, Director of NDRC, Minister of Science and Technology, Minister of Industry and Information Technology, Minister of Security, Minister of Finance, Minister of MEP, Minister of Transport, Minister of Water Resources, Minister of Commerce, Deputy Governor of the People's Bank, Director of State-Owned Assets Supervision and Administration Commission, Director of the General Administration of Taxation, Director of the Safety Supervision Bureau, Chairman of the China Banking Regulatory Commission, Deputy Chief of the People's Liberation Army, and Director of the National Energy Administration (Bo 2010).
6. Many of China's major travel agencies have launched "escape from smog" tours for the public, the destinations of which are often located in southern China, such as Yunnan and Hainan, or countryside areas with relatively better air quality than adjacent big cities.

References

Alford, WP & Liebman, BL 2000, 'Clean air, clean processes? – the struggle over air pollution law in the People's Republic of China', *Hastings Law Journal*, vol. 52, pp. 703–748.

Alford, WP, Weller, RP, Hall, L, Polenske, KR, Shen, YY & Zweig, D 2002, 'The human dimensions of pollution policy implementation: air quality in rural China', *The Human Dimensions of Pollution*, vol. 11, no. 32, pp. 495–513.

Avi-Yonah, RS & Uhlmann, DM 2009, 'Combating global climate change: why a carbon tax is a better response to global warming than cap and trade', *Stanford Environmental Law Journal*, vol. 28, no. 3, pp. 3–50.

Baron, R, Aasrud, A, Sinton, J, Campbell, N, Jiang, K & Zhuang, X 2012, *Policy options for low-carbon power generation in China: designing an emissions trading system for China's electricity sector*, OECD Publishing, Paris.

Bergsager, H & Korppoo, A 2013, *China's state-owned enterprises as climate policy actors: the power and steel sectors*, Nordic Council of Ministers, Copenhagen.

Binder, S & Neumayer, E 2005, 'Environmental pressure group strength and air pollution: an empirical analysis', *Ecological Economics*, vol. 55, no. 4, pp. 527–538.

Bo, ZY 2010, *China's new national energy commission: policy implications*, NUS East Asian Institute, viewed 10 July 2015, <http://www.eai.nus.edu.sg/publications/files/BB504.pdf>.

Chang, J, Leung, DY, Wu, CZ & Yuan, ZH 2003, 'A review on the energy production, consumption, and prospect of renewable energy in China', *Renewable and Sustainable Energy Reviews*, vol. 7, no. 5, pp. 453–468.

China Economic Weekly 2013, *Caizhengbu zhuanjia: juececeng duiyu tanshui renshi bingwei dachang yizhi (Expert of Ministry of Finance: decision-makers have not reached agreement on carbon tax)*, viewed 14 October 2015, <http://www.csfee.org.cn/ReadNews.asp?NewsID=1310>.

China IRN 2007, *Zong touzi 2 wanyi guli minying qiye jinru kezaisheng nengyuan (Total investment 2000 billion huan RMB to encourage private investment into renewable energies)*, viewed 14 August 2015, <http://www.chinairn.com/doc/4080/173769.html>.

Christensen, T, Dong, LS & Painter, M 2008, Administrative reform in China's central government – how much learning from the West?', *International Review of Administrative Sciences*, vol. 74, no. 3, pp. 351–371.

DiPeso, J 2009, 'Carbon tax versus cap-and-trade', *Environmental Quality Management*, vol. 18, no. 4, pp. 95–100.

Downs, ES 2008, 'China's "new" energy administration: China's National Energy Administration will struggle to manage the energy sector effectively', *China Business Review*, vol. 35, no. 6, pp. 42–45.

Dryzek, JS 2013, *The politics of the earth: environmental discourses*, Oxford University Press, Oxford.

Dwivedi, OP & Vajpeyi, DK 1995, *Environmental policies in the third world: a comparative analysis*, Mansell, London.

Ferdinand, P 1991, *Communist regimes in comparative perspective: the evolution of the Soviet, Chinese and Yugoslav models*, Harvester Wheatsheaf, Hemel Hempstead.

General Office of the State Council 2005, *Guwuyuan bangongting tuanyu tuixing xingzheng zhifa zerenzhi de ruoganyijian (General Office of the State Council several suggestions on law enforcement on administrative responsibility system)*, viewed 9 October 2015, <http://www.gov.cn/gongbao/content/2005/content_64332.htm>.

General Office of the State Council 2009, *Guihua huanjing yingxiang pingjia tiaoli (Planning Environmental Impact Assessment Ordinance)*, viewed 7 July 2015, <http://www.gov.cn/zwgk/2009-08/21/content_1398541.htm>.

Goldman, M 1999, 'The emergence of politically independent intellectuals', in M Goldman & R MacFarquhar (eds), *The paradox of China's post-Mao reforms*, Harvard University Press, Cambridge.

Goulder, LH & Schein, A 2013, *Carbon taxes vs. cap and trade: a critical review (No. w19338)*, National Bureau of Economic Research of USA.

Hao, D 2015, *Kezaisheng nengyuan butie quekou 140yi guojia nengyuanju "jinji" modi (Renewable energy subside gap reaches 14 billion yuan, National Energy Administration called for emergency monitoring)*, viewed 15 August 2015, <http://news.xinhuanet.com/energy/2015-04/10/c_1114923913.htm>.

He HH 2015, *Social ethics in a changing China: moral decay or ethical awakening?*, Brookings Institution Press, Washington DC

Heggelund, G 2007, 'China's climate change policy: domestic and international developments', *Asian Perspective*, vol. 31, no. 2, pp. 155–191.

Huang, YS, Ren Wei & Liu, H 2013, 'Wei zhi wumai dizhi shaokao, ni za kan? (Banning barbecue to treat city smog, what is your opinion?)', *Bandao Dushi Bao*, viewed 4 September 2015, <http://bddsb.bandao.cn/data/20130222/html/12/content_1.html>.

Huang, XH 2008, *2008 nian guanyu sheli guojia nengyuanbu he huangbao zongju shengge wei huanbaobu de jianyi (2008 suggestion on establishing the National Energy Administrtion and raising National Environmental Protection Agency to Ministry of Environmental Protection)*, viewed 12 October 2015, <http://www.92office.com/e/action/ShowInfo.php?classid=576&id=12905>.

Huang, S 2014, *Xinnengyuan qiye zaoyu qianhuang (New energy enterprises experience financial difficulties)*, viewed 20 November 2015, <http://energy.people.com.cn/n/2014/0705/c71661-25242441.html>.

Huang, T 2015, 'Qiongding zhixia, shenxian wumai de gonggong yulun (Under the dome, deeply buried public opinions)', *Gongshi Wang*, viewed 2 September 2015, <http://www.21ccom.net/articles/china/gqmq/20150309121953.html>.

Huang, Y 2014, *Zhengfu quanli qingdan zhidu yingnan erjin, xue jiejue hiquan jiaocha deng wenti (Implementation of disclosing power lists mechanism facing difficulty: problems such as overlapping authority need to be fixed)*, viewed 9 October 2015, <http://news.xinhuanet.com/politics/2014-06/09/c_126592385.htm>.

IEA 2011, *World energy outlook 2011*, IEA, Paris.

Jahiel, A 1998, 'The organization of environmental protection in China', *China Quarterly*, vol. 156, pp. 757–787.

Jänicke, M 1996, 'Democracy as a condition for environmental policy success: the importance of non-institutional factors', in W Lafferty & J Meadowcroft (eds), *Democracy and environmental: problems and prospects*, Edward Elgar, Cheltenham, pp. 71–85.

Jing, J 2000, 'Environmental protests in rural China', in EJ Perry & M Selden (eds), *Chinese society: change, conflict and resistance*, Routledge, New York, pp. 143–160.

Kan, H, Chen, B & Hong, C 2009, 'Health impact of outdoor air pollution in China: current knowledge and future research needs', *Environmental Health Perspectives*, vol. 117, no. 5, p. A187.

Keohane, NO 2009, 'Cap and trade, rehabilitated: using tradable permits to control US greenhouse gases', *Review of Environmental Economics and Policy*, vol. 3, no. 1, pp. 42–62.

Knup, E 1997, 'Environmental NGOs in China: an overview', *China Environment Series*, vol. 1, no. 3, pp. 9–15.

Lan, X, Simonis, UE & Dudek, DJ 2006, *Environmental governance in China: report of the Task Force on Environmental Governance to the China Council for International Cooperation on Environment and Development (CCICED)*, Beijing.

Landry, PF 2008, *Decentralised authoritarianism in China: the Communist Party's control of local elites in the Post-Mao Era*, Cambridge University Press, New York.

Lei, J 2014, 'Zhongguo huanbao zuzhi dianming qianjia wuranyiye: jiaju wumai tianqi (Chinese NGOs identify top 1000 polluting companies: who intensify atmospheric pollution)', *Cankao Xiaoxi*, viewed 2 September 2015, <http://china.cankaoxiaoxi.com/2014/1211/593687.shtml>.

Lewis, JI 2008, 'China's strategic priorities in international climate change negotiations', *The Washington Quarterly*, vol. 31, no. 1, pp. 155–174

Li, HY & Wang, Y 2012, 'Yunnan Qujing gezha wuran minjian zuzhi susongan jinxing zhengju jiaohuan (ENGOs conducting evidence exchange for Yunnan Qujing chromium slag pollution case)', *Xinhua Wang*, viewed 9 September 2015, <http://news.xinhuanet.com/local/2012-05/23/c_112022121.htm>.

Li, XW 2007, 'Guanting xiaohuodian sixiang zhuyi (Four issues of closure of small coal-fired power plants)', *CPNN*, viewed 25 February 2016, <http://www.cpnn.com.cn/jdwt3/gd/200704100047.htm>.

Li, M 2014, *Yunnan gezha wuran an – minjian huanbao zuzhi tiqi gongyi susong diyi an (Yunnan chromium slag pollution case – the first environmental lawsuit initiated by grassroots environmental organization)*, viewed 24 June 2015, <http://chuansong.me/n/730440>.

Li, DF 2015, 'Touguo wumai kan huanbao gongyi zuzhi de jianshexing xingdong (Environmental NGOs' innovative actions through city smog)', *Shehui Yu Gongyi*, vol. 4, pp. 34–35.

Liang, JL 2012, 'Fagaiwei guanyuan: tanshui he tanjiaoyi huo erzhezeyi (NDRC official: either carbon tax or carbon trading)', *Economic Information Daily*, viewed 24 July 2015, <http://news.xinhuanet.com/fortune/2012-06/11/c_123263613.htm>.

Liang, JL 2015, 'Zhangzhou PX beihou boyi: huanbaobu shuo guojia fagaiwei "weifa", (Gaming behind Zhangzhou PX: Ministry of Environmental Protection claiming

National Development and Reform Commission "violating the law")', *Sohu News*, viewed 7 July 2015, <http://news.sohu.com/20150412/n411135333.shtml>.

Lieberthal, K 2004, *Governing China: from revolution through reform*, W.W. Norton & Company, New York.

Lin, H & Gil, J 2016, 'Confucian principles to enhance China's environmental governance', in C Andressen (ed), *Changing economies: the interplay between global and Chinese economic change and their impact on Australia*, Routledge, Oxford, pp. 136–157.

Lin, H 2009, A Confucian approach to China's environmental governance a case study of China's atmospheric pollution problem, master thesis, Flinders University.

Lo, AY 2016, *Carbon trading in China: environmental discourse and politics*, Palgrave MacMillan, Basingstoke.

Low Carbon of China 2011, *"Yindui qihou bianhua" shuyuan zhongguo pian: jigou zhengce jiaoliu (Evolution of climate change mitigation in China: organizations, policies and communications)*, viewed 9 October 2015, <http://www.ditan360.com/qihou/qihou_china.aspx?SpecialsID=1138>.

Ma, X & Ortolano, L 2000, *Environmental regulation in China: institutions, enforcement and compliance*, Rowman and Littlefield, Lanham, MD.

Martens, S 2006, 'Public participation with Chinese characteristics: citizen consumers in China's environmental management', *Environmental Politics*, vol. 15, no. 2, pp. 211–230.

McAllister, LK 2009, 'The overallocation problem in cap-and-trade: moving toward stringency', *Columbia Journal of Environmental Law*, vol. 34, no. 2, pp. 395–445.

MEE of PRC 2019, History, viewed 6 February 2019, <http://english.mee.gov.cn/About_MEE/History/>.

MEE of PRC 2019a, Department of Climate Change, viewed 6 February 2019, <http://english.mee.gov.cn/About_MEE/Internal_Departments/201605/t20160526_346894.shtml>.

Mol, APJ & Carter, NT 2006, 'China's environmental governance in transition', *Environmental Politics*, vol. 15, no. 2, pp. 149–170.

Molina, MJ & Molina LT 2012, 'Megacities and atmospheric pollution', *Journal of the Air & Waste Management Association*, vol. 54, pp. 644–680.

Montinola, G, Qian, YY & Weingast, BR 1995, 'Federalism, Chinese style: the political basis for economic success in China', *World Politics*, vol. 48, no. 1, pp. 50–81.

NEA 2012, *Guojia nengyuanju guanyu gouli he yindao minjian ziben jinyibu kuoda nengyuan lingyu touzide shishiyijian, [Guojia Nengyuanju 179 hao] (National Energy Administration's recommendations to encourage and guide private investment in the energy sector, [National Energy Administration No. 179 Document])*, viewed 2 August 2015, <http://www.nea.gov.cn/2012-06/20/c_131665600.htm>.

Ngok, K & Zhu, G 2007, Marketization, globalization and administrative reform in China: a zigzag road to a promising future, *International Review of Administrative Sciences*, vol. 73, no. 2, pp. 217–233.

NDRC 2018, Main functions of the NDRC, viewed 6 February 2019, <http://en.ndrc.gov.cn/mfndrc/>.

Otsuka, K 2002, 'China, social restructuring and the emergence of NGOs', in S Shigetomi (ed), *The state and NGOs: perspective from Asia*, Institute of Southeast Asian Studies, Singapore, pp. 222–244.

Peng, LG & Fang, F 2013, *Sannian dengdai, fagaiwei zhong hezhun; bannian xinxi, huanbaobu zai banghe – zui mingan PX xiangmu huanping weigui shimo (Three years waiting, National Development and Reform Commission finally issued approval; half a year excitement, Ministry of Environmental Protection again got hammered – the complete story of the violation of environmental impact assessment of the most sensitive PX project)*, viewed 13 October 2015, <http://www.infzm.com/content/86048>.

People's Congress of PRC 2003, *Zhonghua renmin gongheguo huanjing yingxiang pingjia fa (Environmental Impact Assessment Law of PRC)*, viewed 12 October 2015, <http://www.gov.cn/gongbao/content/2002/content_61822.htm>.

People's Congress of PRC 2014, *Zhonghua renmin gongheguo huanjing baohu fa (Environmental Protection Law of PRC)*, viewed 13 October 2015, <http://www.npc.gov.cn/huiyi/lfzt/hjbhfxzaca/2014-04/25/content_1861320.htm>.

People's Daily Online 2009, *Huanbao juzhang tong shuo zhifa nan (Minister of Environmental Protection expressed difficulty in environmental law enforcement)*, viewed 3 January 2012, <www.people.com.cn/GB/34948/34951/34953/222739/html>.

Qi, Y & Wu, T 2013, 'The politics of climate change in China', *Wiley Interdisciplinary Reviews: Climate Change*, vol. 4, no. 4, pp. 301–313.

Richerzhagen, C & Scholz, I 2008, 'China's capacities for mitigating climate change', *World Development*, vol. 36, no. 2, pp. 308–324.

Schwartz, J 2004, 'Environmental NGOs in China: roles and limits', *Pacific Affairs*, vol. 77, no. 1, pp. 28–49.

SERC 2011. *Electricity Regulatory Annual Report 2010*, Sate Electricity Regulatory Commission PR China, Beijing.

Shapiro, J 2012, *China's environmental challenges*, Polity Press, Cambridge.

Shapiro, J 2013, 'The evolving tactics of China's green movement', *Current History*, vol. 112, no. 755, pp. 224–229.

Shen, JF, Xue, S, Zeng, M, Wang, Y, Wang, YJ, Liu, XL & Wang, ZJ 2014, 'Low-carbon development strategies for the top five power generation groups during China's 12th Five-Year Plan period', *Renewable and Sustainable Energy Reviews*, vol. 34, pp. 350–360.

Shen, RF 2015, *Tenglong fangting (Zhangzhou) youxian gongsi "4.6" baozha zhaohuo shigu diaocha baogao gongbu (Release of Investigation Report on Dragon Aromatics (Zhangzhou) "4.6" explosion insident)*, viewed 12 October 2015, <http://finance.ifeng.com/a/20150816/13912150_0.shtml>.

Shi, H & Zhang, L 2006, 'China's environmental governance of rapid industrilisation', *Environmental Politics*, vol. 15, no. 20, pp. 271–292.

Shirk, SL 1993, *The political logic of economic reform in China*, University of California Press, Los Angeles.

Song, LF 2009, *Zuixin lishi banben: zhonghua renmin gongheguo guojia fazhan he gaige weiyuanhui (The lasted history: National Development and Reform Commission of PRC), Renmin University Economic Forum*, viewed 18 September 2015, <http://wiki.pinggu.org/edition-view-6054-1.html>.

Song, ZY, Zhang, JX, Wang, YL & Li, J 2008, '30 nian gaige zheshi zhongguo xingzheng guanli tizhi zouxiang (Directions of China administrative system reflected through 30 years of reform)', *Dangde Jianshe*, no. 4, pp. 56–57.

State Council of PRC 2008, *Guowuyuan guanyu jigou shezhi de tongzhi – guofa [2008] 11 hao (State Council Notification on administration framework [2008 State Document no. 11])*, viewed 18 September 2015, <http://www.gov.cn/zhengce/content/2008-04/26/content_7621.htm>.

Sun, JB 2014, *Kezaisheng nengyuan butie jiuzhang nanshou zijin quekou gaoda baiyi (renewable energy subsides owed to industries up to tens of billions yuan)*, CN Energy, viewed 15 August 2015, <http://www.ceh.com.cn/UCM/wwwroot/development/ny/xny/2014/11/777085.shtml>.

Sutter, RG 2008, *Chinese foreign relations: power and policy since the Cold War*, Rowan & Littlefield Publishers, Lanham.

The Central People's Government of PRC 2007, *Guowuyuan guanyu chengli guojia yingdui qihou bianhua ji jieneng jianpai gongzuo lingdao xiaozu de tongzhi, Guofa 18 hao (Notification of the State Council on the establishment of National Leading Group on Climate Change and Energy Saving and Pollution Reduction, Document No. 18)*, viewed 14 September 2015, <http://www.gov.cn/gongbao/content/2007/content_686012.htm>.

Tsui, KY & Wang, Y 2008, 'Decentralization with political trump: vertical control, local accountability and regional disparities in China', *China Economic Review*, vol. 19, no. 1, pp. 18–31.

van Rooij, B 2010, 'The people vs. pollution: understanding citizen action against pollution in China', *Journal of Contemporary China*, vol. 19, no. 63, pp. 55–77.

Walt, G & Gilson, L 1994, 'Reforming the health sector in developing countries: the central role of policy analysis', *Health Policy and Planning*, vol. 9, no. 4, pp. 353–370.

Wang, T, He, HN & Wang, L 2015, *11 geyue qian xiaoshi de gonghan: Huanbaobu jianyi fagaiwei chexiao gulei shihua guihua peifu (A government letter disappeared 11 months ago: Ministry of Environmental Protection suggested National Development and Reform Commission to abolish the approval for Gulei Petrochemical facility)*, viewed 12 October 2015, <http://www.infzm.com/content/108724>.

Wang, Q 2010, 'Effective policies for renewable energy – the example of China's wind power – lessons or China's photovoltaic power', *Renewable and Sustainable Energy Reviews*, vol. 14, no. 2, pp. 701–712.

Wang, Q 2014, 'China should aim for a total cap on emissions', *Nature*, vol. 512, no. 7513, p. 115.

Wang, JH 2015, *Xin huanbaofa zhishao xuyao guo wuguan guanzhu "renxing" de paiwu dahu (The new Environmental Protection Law needs to at pass at least five barriers to control "insubordinate" big polluters)*, viewed 14 October 2015, <http://jjckb.xinhuanet.com/2015-03/16/content_540904.htm>.

Weidner, H 2002, 'Capacity building for ecological modernization: lessons from cross-national research', *American Behavioural Scientist*, vol. 45, no. 9, pp. 1340–1368.

Wen, G 2013, 'Wumai "shengyi": "gongzhong beiju" yanshengde shangji (Smog business: business opportunity emerged from the public tragedy)', *Zhongjing Guancha*, viewed 2 September 2015, <http://www.cb.com.cn/special/show/264.html>.

Williams, L 2014, 'China's climate change policies: drivers and actors', *Lowy Institute for International Policy*, viewed 11 March 2013, <https://www.files.ethz.ch/isn/182715/chinas-climate-change-policies.pdf>.

Wong, E 2015, 'China blocks web access to Under the Dome documentary on pollution', *The New York Times*, viewed 20 January 2018, <https://www.nytimes.com/2015/03/07/world/asia/china-blocks-web-access-to-documentary-on-nations-air-pollution.html>.

Worthley, JA & Tsao, KK 1999, 'Reinventing government in China', *Administration and Society*, vol. 31, no. 5, pp. 571–587.

Wu, FS 2002, 'New partners or old brothers? GONGOs in transnational environmental advocacy in China', *China Environment Series*, vol. 5, pp. 45–58.

Wu, FS 2003, 'Environmental GONGO autonomy: unintended consequences of state strategies in China', *The Good Society*, vol. 12, no. 1, pp. 35–45.

Wübbeke, J 2013, 'China's climate change expert community – principles, mechanisms and influence', *Journal of Contemporary China*, vol. 22, no. 82, pp. 712–731.

Xie, W 2007, 'Zhongguo gonggong zhengce zhixing guocheng zhong de zhengfu bumen jian liyi boyi (Competitions between Chinese government departments in the process of public policy implementation)', *Collected Works of Young Scholars*, vol. 1, pp. 222–228.

Xing, JH 2014, 'Minqi huojiang chengwei xinnengyuan lingyu zhuliju (Private investment may become the main force of the renewable energy sector)', *CPNN*, viewed 15 August 2015, <http://www.cpnn.com.cn/zdgc/201410/t20141022_749259.html>.

Xinhua News 2014, *Zhonghua renmin gongheguo huanjing baohufa zi 2015 nian 1 yue 1 ri qi shishi (Environmental Protection Law of People's Republic of China took effect since 1st January 2015)*, viewed 2 September 2016, <http://www.gov.cn/xinwen/2014-04/25/content_2666328.htm>.

Xinhua News CN Stock 2015, *Kezaisheng nengyuan butie quekou 140yi guojia nengyuanju 'teji' modi (Renewable energy subsidy reaches 14 billion yuan, National Energy Administration emergency investigation)*, viewed 8 September 2015, <http://news.xinhuanet.com/energy/2015-04/10/c_1114923913.htm>.

Xu, CG 2011, 'The fundamental institutions of China's reform and development', *Journal of Economic Literature*, vol. 49, no. 4, pp. 1076–1151.

Yang, D & Xia, XL 2014, *Xu Shaoshi: qunian yilai yi quxiao huo xiafang xingzheng shenpi quan 416 xiang (NDRC has cancelled or given up 416 administrative and approval authority)*, viewed 3 March 2015, <http://finance.people.com.cn/n/2014/0305/c1004-24537680.html>.

Yang, DL 2006, 'Economic transformation and its political discontents in China: authoritarianism, unequal growth, and dilemmas of political development', *Review of Political Sciences*, vol. 9, pp. 143–164.

Yang, GB 2010, 'Brokering environment and health in China: issue entrepreneurs of the public sphere', *Journal of Contemporary China*, vol. 19, no. 63, pp. 101–118.

Zhang, JH 2015, *Changqiong zhixia: huanjing baohu shibuwodai (Thoughts on Under the Dome: the urgency of environmental protection)*, viewed 9 September 2015, <http://news.xiancity.cn/readnews.php?id=206684>.

Zhang, KJ 2007, 'The shadow of environmental pollution in China', in CJ Liang & DP Yang (eds), *The China environmental yearbook (2005): crisis and breakthrough of China's environment*, Social Science Academic Press, Beijing, pp. 179–196.

Zhang, WT & Zhang PR 2015, *Chen Jining: jiang tuichu weixin jubao wuran, chachu weipi xianjian weifa qiye (Chen Jining: to establish WeChat pollution reporting platform and investigate and punish enterprises established without gaining approval)*, viewed 14 October 2015, <http://env.people.com.cn/n/2015/0307/c1010-26654312.html>.

Zhao, H 2015, *Qiongding zhixia, zaibeijing yu wumai sikede 10jia huanbao zuzhi (Under the Dome, 10 environmental NGOs that set dead on air pollution in Beijing)*, viewed 2 September 2015, <http://www.yogeev.com/article/56976.html>.

Zhao, ZY, Zuo, J, Fan, LL & Zillante, G 2011, 'Impacts of renewable energy regulations on the structure of power generation in China – a critical analysis', *Renewable Energy*, vol. 36, no. 1, pp. 24–30.

Zhi, H 2013, *Wuranwu paifang shi zaocheng wumai tianqi de "zuikuihuoshou" (Pollutants discharge is the cause of city smog)*, viewed 4 September 2015, <http://www.weather.com.cn/index/zjzx/01/1788533.shtml>.

Zhihu 2015, *'Boda 12369 huanbao rexian youyongma? (Does it work to dial 12369 environmental protection hotline?)'*, viewed 10 May 2017, <https://www.zhihu.com/question/28437244>.

Zhu, XF 2009, 'The influence of think tanks in the Chinese policy process: different ways and mechanisms', *Asian Survey*, vol. 49, no. 2, pp. 333–357.

Zhu, XF & Xue, L 2007, 'Think tanks in transitional China', *Public Administration and Development*, vol. 27, no. 5, pp. 452–464.

4 From CDM to pilot carbon trading scheme

Introduction

In order to simultaneously fulfil the tasks of mitigation for global climate change and sustainable development in developing countries, the United Nations Framework Convention on Climate Change (UNFCCC) proposed the concepts of cost-effective emission mitigation and flexible cross-region cooperation (Dutschke & Michaelowa 1998, p. 7). The Clean Development Mechanism (CDM) was one such flexible market mechanism tailored for the above concepts under the Kyoto Protocol. It aimed to reduce emission reductions costs in Annex I developed countries and to promote sustainable development in non-Annex I developing countries. CDM allowed an Annex I country to invest in low-cost emission reduction projects in developing countries which did not have quantified emission targets under the Kyoto Protocol. By doing so, the Annex I investor could obtain Certified Emission Reductions (CERs) (1 CER = 1 metric ton of CO_2 or CO_2 equivalent [tCO_2e]) to count towards its total emission reduction obligation. Meanwhile, the projects should also be beneficial for the host country's sustainable development (Article 12, Kyoto Protocol).

Sustainable development is China's long-term national strategy. A range of national policies aiming to enhance sustainable development have been introduced over the years. Policies on energy conservation, environmental protection, and reforestation, for example, have all contributed to sustainable development and simultaneously brought positive impacts in combating climate change. The ratification of the Kyoto Protocol in 2002 gained China eligibility to participate in CDM, and China gradually became a key player in this scheme since then (The World Bank et al. 2004, pp. xix-xx). So far, China has hosted the largest number of CDM projects among all developing nations. Up until 5 May 2015, 5073 CDM projects had been approved nationally. By 15 June 2015, 1450 Chinese-hosted CDM projects had been issued CERs by the CDM Executive Board (CDM EB), which accounted for over 60% of global total CERs and an annual average reduction of 344 million tCO_2e (Department of Climate Change, NDRC 2015).

Following the success of CDM, China actively researched climate change mitigation through market-based instruments. In October 2011, National Development and Reform Commission (NDRC) announced the official launch of China's pilot carbon trading program. Five municipal cities including Beijing, Shanghai, Tianjin, Shenzhen, and Chongqing, and two provinces Hubei and Guangdong were chosen for the pilot program.

This chapter will explore what CDM means to China's sustainable development. It will also evaluate the strengths and weaknesses of China's CDM program and pilot carbon trading program. Both programs are the initial attempts for China to conduct climate change mitigation through market means. To understand what they mean and their strengths and weaknesses will bring useful insights for China's future climate policies.

China's participation in international CDM

CDM was the only major international cooperation on climate change mitigation China participated in during the 11th and 12th Five-Year Plans. The development of CDM in China began slowly. By the end of 2005, there were only three registered CDM projects nationwide (Maraseni & Gao 2011, p. 340). Apart from the complex and time-consuming nature of working through the CDM legal framework and administrative system, this situation was also held up by a concern that CDM might curtail China's cheap climate change abatement options, the so-called "low-hanging fruit". The reduction of cheap abatement options might leave future abatement harder and more costly (Castro 2012, p. 199). However, after the initial hesitation, China gradually fully embraced CDM.

This shift in attitude towards CDM was due to a range of reasons. First of all, CDM was a market mechanism. The high demand for CERs was undoubtedly the biggest drive for China's attitude shift. After late 2005, CDM experienced a boom in China. This boom was primarily driven by the high demand and rising price of the European Union (EU) emission allowance (the primary carbon credits in EU emission trading scheme [ETS]), in which prices reached a high of €30 (approximately US$36) in July 2005. The price of CERs were below that of the EU emission allowance as CERs were generated from cheaper abatement options in developing countries. CERs could be used the same way as the EU allowance by Annex I countries to comply with their emission targets. Therefore, CDM projects appeared feasible and attractive to investors (Michaelowa & Buen 2012, p. 5). On top of this, the industrial gas projects, such as Trifluoromethane (HFC-23) and N_2O, had further inflamed the boom and pushed forward the development of CDM in China due to their highly profitable nature. The technologies incorporated for the reduction of industrial gases were relatively inexpensive, but warming effects of these industrial gases were much higher than CO_2. For instance, the warming effect of HFC-23 is 11,700 times of CO_2. This is to say, by reducing only one unit of HFC-23, the investor would obtain 11,700 CERs.

This also meant that CERs from industrial gas projects were among the cheapest types of all CDM projects (Michaelowa 2012, p. 19).

The majority of China's industrial gas projects were registered in the early stage of China's CDM involvement. According to Maraseni and Gao (2011, p. 342), up to 2010, HFC-related projects were only 0.5% of China's total CDM projects, but they accounted for over 14% of expected CDM annual emission reduction. N_2O-related projects were not far behind by generating about 4% of total CDM emission reduction while only taking up 1% in proportion of total CDM projects.

Although industrial gas projects accounted for a proportionally large percentage of greenhouse gases (GHG) emission reduction (CO_2e), they did not significantly contribute to local sustainable development. First of all, industrial gas projects did not assist energy structure optimisation in host countries because they did not tackle the most dominant and long-lasting greenhouse gas – CO_2. Therefore, they had a weak impact on the fundamental challenge of China's climate change mitigation – how to reduce reliance on fossil fuels and how to switch to a mix of low carbon renewable energy sources (Pearson 2007, p. 247). Industrial gas projects, although they reduced GHG emissions, did not force systemic change. Second, the majority of industrial gas projects happened on the sites of existing facilities, and had little contribution to the improvement of local infrastructure and employment. As a result, the integrity of industrial gas projects on sustainability had been questioned by many (Pearson 2007; Schneider 2007; Schwank 2004). Furthermore, the EU no longer recognised CERs generated from industrial gas projects after January 2013 (Gtowacki Law Firm 2013).

The Chinese government realised this limitation and shifted its investment focus from industrial gas projects to renewable energy projects. In the revised CDM regulation *Measures for Operation and Management of Clean Development Mechanism Projects in China* effected on 12 October 2005, Article 4 identified that China's priority CDM projects were in the category of renewable energy, energy efficiency improvement, and methane recovery and utilisation; Article 24 stated that all Chinese CDM projects were subjected to a tax on their CER sales. Projects in the priority categories would be charged a tax of 2%. Industrial gas projects such as HFC and perfluorocarbon PFC projects were charged a much heavier 60% tax, and N_2O projects were charged at 30% (NDRC 2005, n.p.). *China Medium and Long Term Energy Conservation Plan* also stated that the tax collected would go to a special fund used only for climate-related activities, which was managed by Ministry of Finance (MF) and jointly decided by NDRC and all relevant ministries (NDRC 2004a, n.p.). Since then, the investment direction of China's CDM projects had been successfully redirected towards the priority categories including renewable energy projects, energy saving and efficiency improvement projects, and methane recovery projects. The first two types, which focused on China's energy structure optimisation and energy conservation, tackled the most problematic greenhouse gas CO_2, and therefore had significant meaning in China's long-term climate change mitigation (Wang & Chen 2010, p. 1996). For this reason, this chapter focuses on the analyses of these two types of CDM projects, renewable energy projects and energy saving and efficiency improvement projects.

Institutional capacity building and approval process of Chinese CDM projects

In 2004, NDRC established the Designated National Authority (DNA) catered for China's CDM governance. Any project aiming to register as a CDM project had to first submit its application to the DNA. The application was then forwarded to an expert board for initial assessment. If the application passed the initial assessment, it would be passed on to CDM EB for the final decision. If the application was rejected by CDM EB, it would have a chance to revise and resubmit (Heggelund 2007, p. 181). Within the central government, NDRC and Ministry of Science and Technology (MOST) co-chaired the National CDM Board. NDRC was responsible for the policies and overall governance of CDM. MOST co-supervised CDM projects by providing science and technology research on CDM. Both ministries followed the coordination of the National Leading Group on Climate Change (NLGCC), which was the top supervising authority of China's climate change affairs (Schröder 2012, p. 49). At the provincial level, CDM management centres were set up. These centres were responsible for providing detailed supervision and assistance to CDM projects; they were also responsible for maintaining CDM data, organising CDM experts, and reporting information to the Department of Climate Change (DCC), NDRC (Heggelund 2007, p. 181).

The DNA's approval process was regarded as fairly transparent and the system appeared vigorous and reliable. Priority projects were mostly handled in a hassle-free and prompt manner (Heggelund 2007, p. 181). However, as the CDM EB's approval criteria changed and toughened over the years, there were some Chinese projects which passed NDRC's initial assessment but failed to meet CDM EB's criteria. As a result, they were unable to register as CDM projects. Passing the NDRC's assessment indicated that the projects were proposed and planned according to the relevant CDM methodology, and had obtained approval for development in China. But failure to register with CDM EB meant the projects could not realise their CER revenue, which would lead to financial pressure of the applicants, especially those with high investment requirements and low returns. A proportion of China's renewable energy CDM projects ended up in this category. The situation was partially due to NDRC's unprompted adaptation to EB's approving criteria updates, leaving some applications unrevised before submission. It could also be caused by the developers' uncomprehensive application documentation and the negligence of the expert board at the initial assessment. The success rate of CDM registration could have been improved through institutional capacity building. For instance, NDRC could improve its pace for adapting EB policy updates. NDRC could also enrich expertise and resources at CDM management centres to provide better assistance for applicants. Furthermore, NDRC should have established tighter criteria on additionality for initial assessment. Among all these possible improvements, the criteria on projects' additionality were of great importance. This was because EB did not approve projects which would have happened anyway in the business as usual scenario.

Additionality test of CDM projects

Under the Kyoto Protocol, emission reductions from CDM projects were required to be real, measurable, and additional. Being additional was the key criteria for a project to gain CDM approval, which meant that a project and its associated emission reduction would have not occurred without the registration of CDM (Schneider 2009, p. 243). For example, a Chinese City "A" needs to build a new power-generating facility to cope with the increasing energy demand. It has coal and wind resources available. Under the conventional circumstances, a coal-fired power plant is more likely to be considered as it is a cheaper option. However, with the registration of CDM, the revenue from CERs sale will compensate part of the investment cost of a wind farm, which makes the wind farm option more feasible. Therefore, building a wind farm avoided the emissions from the coal-fired power plant. By this means, the wind farm in City "A" is additional. To the contrary, City "B" also needs to build a new power facility which only has wind resources but no access to coal or other fossil-based resources due to remote location and poor road transport. So, it builds a wind farm. The wind farm in City "B" cannot be classified as additional as it is the only option for the city and would have happened anyway. Therefore, it would not qualify for CDM registration. The identification of a project's additionality is important because, as an offsetting mechanism, it needs to prevent double counting of global emission reduction (Michaelowa & Buen 2012, p. 19). If the wind farm in City "B" is registered as a CDM project, then every single CER it sells would mean an extra ton of CO_2e released into the atmosphere by the CER buyer. Under China's current situation, building coal-fired power plants is still the cheapest option in the majority of locations due to abundant coal reserves and an advanced rail and highway transport systems (Lin & Liu 2010). Therefore, the majority of renewable energy projects approved by NDRC where coal resources are accessible should be regarded as additional.

There were a range of additionality assessment methodologies that had been developed over the years. The "investment analysis", the "common practice analysis", and the "positive lists" analysis were among the most commonly used methods for assessing the additionality of the proposed renewable energy projects and energy saving and efficiency improvement projects. The barrier analysis needs to demonstrate that barriers exist that would prevent the proposed project from being carried out if not registered as a CDM project. The investment analysis needs to demonstrate that the proposed project is economically or financially less attractive than at least one other credible alternative. The common practice analysis requires an assessment of the extent to which the proposed project type (e.g., technology or practice) has already been deployed in the relevant sector and region. (Schneider 2009, pp. 243–244). The CDM EB could either combine barrier and investment analysis or use them separately to assess the additionality of large scale projects, and then use common practice analysis as a credibility check. For small scale projects, CDM EB would only perform a simple barrier test (Schneider 2009, p. 244). These additionality assessment methodologies were important as they not only provided guidance for NDRC to update its rules and policies over the initial assessment of Chinese projects, but they also guarded the integrity of Chinese CDM projects in fulfilling the task of global emission reduction.

Different approaches to CDM projects under Kyoto Protocol

Under the Kyoto Protocol, Article 12 allowed a range of flexible approaches to CDM projects, including the bilateral approach, the multilateral approach, and the unilateral approach. The bilateral approach involved both Annex I developed countries and non-Annex I developing countries (host countries). In the multilateral approach, an intermediary would arrange a set of desired complementary CDM activities for its clients. The unilateral approach refers to CDM projects solely carried out by non-Annex I countries without direct involvement of Annex I partners in the planning, financing, and implementing processes (Belis 2013, p. 162). The bilateral and multilateral approaches were initially desired during the negotiation of CDM under Kyoto as it would lower the cost of Annex I countries' emission mitigation and, at the same time, encourage the transfer of modern climate-friendly technologies to non-Annex I developing countries and thus contribute to their sustainable development (Michaelowa 2007, p. 17). The majority of China's bilateral projects were concentrated at China's initial participating period. Later, unilaterally financed projects started to flourish in China. For these unilateral CDM (uCDM) projects, Annex I counties were only involved in the purchase of CERs once the projects were under production, though some might provide assistance to the registration of the projects they were interested in buying CERs from (Greiner 2009, p. 681).

It was commonly agreed upon that uCDM projects had both advantages and risks (Haites, Duan & Seres 2006; Michaelowa 2007; Seres, Haites & Murphy 2009). The attractive features of uCDM projects were: (1) ability to quickly develop a range of small scale projects with sustainability feature; (2) lower transaction cost due to fewer institutional bundles involved compared with bilateral projects; (3) the host country had total control of CERs (Haites, Duan & Seres 2006). This type of projects also presented a range of risks. The most obvious and widely mentioned disadvantage of uCDM was lack of technology transfer. However, the greatest risk for the host countries associated with uCDM was finance. First of all, the host country had to risk premium for potential foreign CER buyers, which required large amount of upfront investment capital. Second, it was a long process from project planning to production and to CER sale; therefore, the financial flow was slow. Third, the potential CER revenue was not certain as it subjected to the fluctuation of global carbon market. Finally, there was the possibility for uCDM projects to not find CER buyers. Under this circumstance, CER revenue could not be realised (Michaelowa 2007, p. 17). Maraseni and Gao (2011) conducted a survey with a group of key Chinese CDM stakeholders on China's uCDM. The results echoed the above statements. Understanding the difficulties and risks of uCDM projects had a significant meaning for the development of China's future small-scale climate change mitigation projects. Many small uCDM projects proposed by small investors and private entities were of great effect on local sustainability development. If the financial obstacles removed and CER sales ensured, this type of small-scale projects might flourish. This would as a result bring greater benefit to the local environment and encourage more private investment into climate change mitigation.

Emission mitigation capacity of China's major energy-related CDM projects (approved by EB)

The carbon emission reduction capacity of China's energy-related CDM projects was significant. Among all EB registered Chinese CDM projects, renewable energy projects and energy saving and efficiency improvement projects took up 90.12% of all types of projects. Renewable energy projects accounted for the biggest average annual reduction of 399,980,124 tCO_2e. Energy saving and efficiency improvement projects generated an average annual reduction of 52,289,555 tCO_2e (Table 4.1).

Table 4.1 Expected average annual CERs by scope (Projects registered by EB up to 14 July 2015) (Unit: tCO_2e = 1 ton of CO_2 equivalent)

Scopes	*Average Annual Reductions*	*Scopes*	*Average Annual Reductions*	*Scopes*	*Average Annual Reductions*
Energy saving and efficiency improvement	52,289,555	Renewable energy	399,980,124	Fuels substitute	18,546,998
Methane recovery and utilisation	51,818,507	N_2O decomposition	28,181,743	Chemical pollutants reduction (HFC-23)	66,798,446
Landfill burning power generation	5,913,492	Afforestation and reforestation	156,486	Others	4,568,326

Source: NDRC (2015).

The combined average annual reduction of the above two types of Chinese CDM projects reached approximately 0.4 billion tCO_2e (NDRC 2015). This was a significant amount as it compared well with China's annual domestic effort in emission reduction. During the 11th Five-Year Plan (2006–2010), China's energy intensity dropped by 19.06% from the 2005 level. This resulted in a total reduction of 1.5 billion tCO_2e (approximately 0.3 billion tCO_2e reduction per year) (State Council of PRC 2012, n.p.). China's planned next step is to achieve a total reduction of 5.5 billion tons of CO_2 during the period 2011–2020 (Xuan & Xue 2012, p. 18). Spreading this amount over 10 years, an average of 0.55 billion tCO_2e emission reduction is required annually. The predicted annual combined CO_2e emission from the two types of energy-related CDM projects (0.4 billion tCO_2e) was higher than China's annual achievement during the 11th Five-Year Plan from domestic effort (0.3 billion tCO_2e per annual) and is not far behind China's new domestic annual reduction target (0.55 billion tCO_2e per annual). This was to say, the two types of China's energy-related CDM projects alone had a great capacity in assisting global emission reduction, which almost equalled China's total domestic mitigation effort. Thus, CDM's effectiveness as a flexible offsetting mechanism should not be overlooked.

The carbon emission reduction capacity of China's renewable energy and energy saving and efficiency improvement CDM projects was high; however, their associated emission reduction achievements did not count as China's effort. This was because CERs from China's CDM projects were sold to Annex I entities and were therefore counted as the buyers' emission reduction. China as a host country solely enjoyed the benefits that CDM projects brought in terms of enhancing sustainable development, such as better local air quality and more efficient and advanced local infrastructures. It is important to make clear the right of attribution of emission reduction achievements of CDM projects so that double counting in total global carbon emission reduction can be avoided. Thus, if the discussion was only oriented at the contribution of CDM towards China's total emission reduction target, the conclusion was clear that CDM did not boost China's emission reduction portfolio.

Projects that potentially boost China's emission mitigation capacity

As mentioned earlier, two types of projects could potentially contribute to China's domestic climate change mitigation if they were absorbed within the Chinese domestic carbon market. They were (1) projects which gained NDRC approval but did not pass EB assessment and (2) unilateral projects which passed EB assessment but had not yet found international CER buyers. According to NDRC data, there were 630 NDRC-approved energy saving and efficiency improvement projects, but only 254 successfully registered with EB. Among 3736 NDRC-approved renewable energy projects, 3177 were qualified for EB assessment criteria. This was to say, 559 renewable

Table 4.2 Difference in number of NDRC-approved projects and EB-registered projects

Scopes	*Project Number: Approved by NDRC/ Registered with EB*	*Scopes*	*Project Number: Approved by NDRC/ Registered with EB*	*Scopes*	*Project Number: Approved by NDRC/ Registered with EB*
Energy saving and efficiency improvement	630/254	Renewable energy	3736/3177	Fuels substitute	49/26
Methane recovery and utilisation	474/235	N2O decomposition	43/43	Chemical pollutants reduction (HFC-23)	11/11
Landfill burning power generation	56/35	Afforestation and reforestation	5/4	Others	68/21

Source: NDRC (2015a)

energy projects and 376 energy saving and efficiency improvement projects with sustainability enhanced features could be absorbed within China's domestic carbon market (Table 4.2).

As demonstrated in Table 4.3, the difference in average annual reductions between NDRC-approved projects and EB-registered projects was also significant. For NDRC-approved energy saving and efficiency improvement

Table 4.3 Difference in average annual reductions between NDRC-approved projects and EB-registered projects (Unit: tCO_2e)

Scopes	*Average Annual Reductions: NDRC/EB*	*Scopes*	*Average Annual Reductions: NDRC/EB*	*Scopes*	*Average Annual Reductions: NDRC/EB*
Energy saving and efficiency improvement	98,363,384/ 52,289,555	Renewable energy	461,944,425/ 399,980,124	Fuels substitute	24,875,953/ 18,546,998
Methane recovery and utilisation	81,836,570/ 51,818,507	N_2O decomposition	28,181,743/ 28,181,743	Chemical pollutants reduction (HFC-23)	66,798,446/ 66,798,446
Landfill burning power generation	9,040,012/ 5,913,492	Afforestation and reforestation	157,610/ 156,486	Others	11,243,941/ 4,568,326

Source: NDRC (2015)

projects which failed to register with CDM EB, there was an emission reduction capacity of 46,073,829 tCO_2e per year, and the annual emission reduction capacities for such type of renewable energy projects was 61,964,301 tCO_2e. If these projects were implemented, the combined average annual reductions would reach 108,038,230 tCO_2e, which was just under a quarter of the total EB-approved Chinese CERs generated from both efficiency improvement and renewable energy projects (462,269,679 tCO_2e). According to the interview conducted with an NDRC policy officer in 2013, NDRC were considering using CERs from such kind of projects in China's pilot carbon trading program. In fact, they were incorporated into the pilot program at the end of the 12th Five-Year Plan. And they definitely have their role in the current national ETS.

There was another type of CDM projects that had the potential to contribute to China's domestic emission reductions at the time. These were unilateral CDM projects that had successfully registered with CDM EB but had not yet found CER buyers. The CERs generated from this type of projects could be traded on the global carbon market. However, NDRC also considered the utilisation of such kind of CERs in the domestic carbon market. Interviewee 1 explained the preparation of China's own carbon offsetting mechanism in October 2013 as below:

> We are now discussing whether we should domestically absorb these types of CERs. NDRC is considering bringing these CERs to domestic carbon market after 2012. As long as the procedure is according to CDM methodology and is approved by NDRC, even though it is not approved by EB or approved by EB but cannot find buyers, these CERs can be traded in the domestic carbon market. And I think it is a very promising aspect. Although the detailed method for operation such as where it can be used among the seven [carbon trading] pilot programs and how it can be used in future national carbon market are not decided yet, the government will definitely recognize and accept these CERs. This is because you [the investors] have done the good things for the environment and the government won't make your effort in vain.

To sum up, CDM, as a global-scale flexible market mechanism, mainly reduced GHG emission in Annex I developed countries. However, two types of projects could potentially boost China's domestic climate change outcomes. These were projects which had been approved by NDRC but had not passed EB assessment and CDM projects successfully registered with EB but failed to find CER buyers. The mitigation capacity of these two types of projects was significant (more than 0.1 billion tCO_2e per annum) when considering China's total mitigation target of 0.55 billion tCO_2e per annum. However, there might be concerns over their additionality even if they were used as a national offsetting mechanism. Some implications will be provided in Chapter 6 in terms of project additionality. In the following section though, the most significant feature of CDM from developing countries' point of view, the promotion of sustainable development, will be discussed in China's context.

The strengths of China's CDM program and contribution for sustainability

Whether CDM promoted sustainable development, or to what extent it did, had attracted much attention. The conclusions varied depending on different conditions of the host countries and different criteria used to evaluate this (Lloyd & Subbarao 2009; Nussbaumer 2009; Olsen & Fenhann 2008; Schneider, Holzer & Hoffmann 2008; Sutter & Parreño 2007). It is argued here that CDM did promote sustainable development in China in the following aspects: raising climate change awareness, promoting development of renewable energies, and providing inspiration for China's domestic carbon market.

Raising awareness of climate change

CDM was of great significance in lifting people's awareness of climate change and how to deal with climate change in a scientific manner. In Schneider's report which evaluated whether CDM was fulfilling its environmental and sustainability objectives, the author concluded that "One of the key contributions of the CDM is its impact on the awareness and understanding about clean technologies, emission trading and future action for climate change both in the private and public sector" (Schneider 2007, p. 73). China, as the host country of the largest number of CDM projects, had benefited greatly from this aspect. NDRC official Interviewee 1 explained this important impact of CDM as follows:

> The greatest benefit of CDM is to let Chinese people, and especially the Chinese companies, know about climate change, and they can benefit from emission reduction. It contributes greatly to the improvement of public awareness on climate change. Take the 11th Five-Year Plan for example, we have achieved a lot in energy saving and pollution reduction, and public awareness in energy conservation has improved a lot. This is because we gave publicity to environmental protection and energy conservation for the last 30 years, and people's awareness improved gradually. But climate change is an urgent issue, we don't have another 30 years to build up the awareness, and CDM did the work.

In fact, signs of the positive impact of CDM on public awareness already showed in the business sector. For instance, Chinese entities which participated in CDM had adopted a set of more standardised operation systems in terms of carbon measurement and auditing. Interviewee 6 from CDM

Non-governmental organizations (NGO) Quanlian New Energy Chamber of Commerce stated:

> The operation process of CDM itself is a systematic operation process for a company, just like the ISO9000.[1] Many small enterprises in China do not have a standardised operation system, but if they have operated under CDM, their operation such as data monitoring and shifts and other procedures will be more standardised. This is very helpful for the company.

From this we can see that the ongoing implementation of CDM had a positive impact on business awareness by providing learning opportunities to Chinese companies for more advanced and standardised operation procedures when dealing with climate change.

A national telephone survey of 4169 Chinese adults, conducted from July to August 2012 by the China Centre for Climate Change Communication, echoed the statements of the above interviewees. The survey focused on investigating Chinese people's awareness, beliefs, attitudes, policy support, and environmental behaviours towards climate change. The results showed that majority of the Chinese public were aware of climate change, believed human activities contribute to climate change and wished the government to pay more attention to climate change (Table 4.4).

This was a positive outcome regarding the Chinese public's climate change awareness, given China's traditional reluctant attitude towards climate change mitigation. It was also a significant improvement compared with the

Table 4.4 National survey of the Chinese public's awareness on climate change

Factor	*Main Results*
Awareness	93% of respondents say they know at least a little about climate change. 11% say they know a lot, 54% know something, and 28% know just a little about it. 7% have never heard of climate change.
Belief	55% say that climate change is caused mostly by human activities, while 38% say that climate change is caused mostly by natural changes in the environment.
Attitude	78% of respondents say they are either very (23%) or somewhat worried (55%) about climate change. 14% are not very worried and 8 percent are not at all worried.
Experience	69% say that people in China are already being harmed by climate change, while another 8% say they will be harmed within 10 years.
Responsibility	89% of respondents agree and 9% somewhat agree with the statement: "The government should pay great attention to the issue of climate change".
Source of information	Majorities of respondents trust scientific institutes (89%), the government (86%), the news media (82%), and their own friends and family members (64%) as sources of information about climate change. Fewer trust NGOs (41%) or corporations (38%).

Source: China Centre for Climate Change Communication (2012)

survey conducted on the same issue by Lin and Gil in 2009 (Lin & Gil 2016). This improvement in people's climate change awareness would have positive impact on public understanding and gain public support for China's future climate change mitigation plans.

Promoting China's renewable energy development

The ability of CDM in promoting renewable energy development was one of its most significant contributions towards sustainable development. Pearson (2007, p. 247) boldly stated that "The question of whether the CDM is promoting sustainable development can be framed primarily in terms of whether it is promoting renewables in developing countries and thus assisting in the transition away from fossil fuel". Boyd et al. (2009) evaluated this matter through an in-depth analysis of 10 CDM projects' sustainability benefits. They concluded that CDM only enhances sustainability to a high extent by installing small-scale renewable energy in remote regions of developing countries to cover the newly emerged or growing energy demands. Regardless of the different criteria for defining sustainable development, it is most commonly agreed upon that the development of renewable energy in appropriate regions with a purpose of replacing or avoiding traditional fossil fuel-based energy has a positive impact on sustainable development.

New and renewable energy-related CDM projects account for 83.5% of China's total approved CDM projects. Up until 14 July 2015, among all 3806 EB-registered CDM projects, 3177 were in the category of new and renewable energy projects (NDRC 2015a, n.p.). China registered more than 40 GW of wind power and 30 GW of hydropower capacity under CDM, which was larger than the total combined Australian national grid (Scotney et al. 2012, p. 7). Within the category of renewable energy, wind power was a new star after China's traditional hydropower and attracted much attention. Solar power, with the greatest development potential, also experienced rapid development with the stimulation of CDM.

Wind power in China experienced a slow development in the initial stage and experienced a boom since the mid-2000s. The early wind power projects in China were often very small in scale in remote areas and were not connected to the grid. Since the mid-1990s, the wind power and related manufacturing industries emerged after the initial identification of domestic regulation detailing wind power installation target and the availability of relevant funding. However, the wind turbines installed were mainly imported as a result of foreign manufacturers' internationalisation plans. Even though foreign manufacturers could provide grants or loans for setting up wind power plants in China, the enthusiasm on the Chinese side was still low. This was due to a combination of reasons including lack of a unified pricing mechanism, banks' unwillingness to lend, high transaction costs and investment risk, and fear of unpopularity of high electricity prices among the public (Buen & Castro 2012, pp. 64–66). The development of China's wind power was underestimated by the authorities for quite a long period. For instance, the Research Team of China Climate Change Country Study (RTCCCCS) (1999) predicted that China's wind power installation capacity would be 2 GW by 2020; while in 2010 China's wind power capacity had already reached 44.7 GW (Global Wind Energy Council 2011, p. 30). During the 11th Five-Year Plan (2006–2010), China's wind power experienced an average annual growth rate at 105%, which is seven times higher than the growth rate of the nation's total energy capacity. By 2010, China's wind power generation was 32 times higher than that of 2005 (Kang et al. 2012, p. 1909). China's global share of wind power production also raised dramatically during this period from 2.1% in 2005 to 22.4% in 2010 with nearly half of the global newly added capacity from China. China became the biggest wind power producer in the world (Chinese Wind Energy Association (CWEA) 2010). The sudden boom of China's wind power and related manufacturing industry after 2003 was led by a range of supportive domestic policies and triggered by the introduction of CDM (Buen & Castro 2012, p. 73).

The first step that the Chinese government took towards a successful wind power industry was to establish a market for wind power products and to protect domestic wind turbine manufacturers and wind power developers from competition. Specifically, it provided concessions for development of large-scale wind farms; provided loans and tax benefits; required developers to purchase 50%–70% of equipment from the domestic manufacturers; made sure that energy generated from wind farms was purchased at an optimal price; and issued long-term power purchasing agreements for 25 years (Buen & Castro 2012, p. 70). All these measures set up a solid foundation for the prompt development of China's wind power industry. In 2009, the revised version of the *Renewable Energy Law* further confirmed that it was the utility companies' obligations to accept renewable energies into their grids, and renewable energies must be purchased continually throughout the power plant's lifetime. The additional cost of renewable energy was to be shared among all consumers nationwide instead of by local consumers where renewable energy was generated. Furthermore, the national grid was required to accept renewable energy generated from regional areas where power could not be completely consumed locally (The National People's Congress of PRC 2013, n.p.). Such guarantees from the revised *Renewable Energy Law* further enhanced the robustness of the wind power industry. Take Gansu Province, for example, the grid connection of wind power increased significantly since the implementation of the new *Renewable Energy Law*. In 2013, the total amount of electricity generated from wind power in Gansu increased 29.73% compared with previous year.

Among all renewable energy CDM projects, the Chinese government gave high priority to wind power projects. This was a strategic plan for China's relatively dry northern and western regions where wind resources were rich but water resources were scarce. Such conditions limited the development of energies that required large amount of water for production. Hydropower, which was the cheapest renewable energy option in many locations of China, required both abundant water and suitable geology. Generation III nuclear power plants also required large amounts of water as coolant. Therefore, wind power became the clear choice for China's *Developing the West and Revitalising the Northeast plans*. Also, wind power had price and technological advantages. It was among one of the cheapest renewable energies, and the technology was relatively mature (Wang & Chen 2010, p. 1993). For large wind power projects that passed EB approval, their CER sales were estimated at about 10% of their total revenue. It therefore appeared attractive to investors (Castro et al. 2011, p. 221).

Solar energy is a renewable energy of the greatest potential for satisfying the world's future energy demand. According to the International Energy Agency's (IEA) prediction, solar power will account for over a quarter of global energy generation in 2050. Photovoltaic (PV) solar will contribute 16% of the world's electricity needs and solar thermal will provide 11% (Evans 2014, n.p.). China's solar resource can be described as super abundant compared with other renewable resources such as wind and hydro resources. There are approximately 100 billion kW solar resources that can be utilised in China, while the total available wind resource is only 1 billion kW and economical hydro resource 0.4 billion kW (Yin et al. 2012, p. 166). Therefore, the research and development of solar energy has a significant role in satisfying China's long-term energy demand and climate change mitigation. China became the leading player in the world's PV solar market since 2013, followed by Japan and the USA. China is expected to continue to lead the global market, accounting for about 37% of global capacity by 2050 (IEA 2014, p. 5). The biggest problem with PV solar is the high requirement of upfront investment; thus, the primary task in the development of PV solar energy is to lower the upfront cost (IEA 2014, p. 1). As technology matures, the price of PV solar energy will drop. According to China's *Medium and Long Term Energy Development Plan*, the generation cost of PV solar electricity will decline to 0.6 yuan/kWh in 2020, which will be very close to the price of coal-generated electricity (Yin et al. 2012, p. 190).

China's solar thermal was mainly concentrated in the utilisation of solar hot water systems during the 11th and 12th Five-Year Plans. So far, China has become the biggest manufacturer of solar hot water systems. Solar hot water systems have already been commercialised in China without any government grants and subsidies. China's next step is to push forward compulsory installation of solar hot water systems for buildings under 20 levels through legislation. The government will also provide incentives for manufacturers to speed up the research and development of solar thermal utilisation (Yin et al. 2012, p. 214).

Nonetheless, the road for China's renewable energy CDM projects was not completely smooth. As a market mechanism, CDM was unavoidably affected by the 2008 global financial crisis and CER price dropped constantly afterwards. Although it received both domestic and international support, China's renewable energy industry was still impacted by the circumstance. When being asked about the impact of the CER price tumble on China's renewable energy CDM projects, Interviewee 6 from Quanlian New Energy Chamber of Commerce stated:

> Renewable energy is a kind of investment associated with high cost and relatively low return, but with CDM, the CERs will compensate some of its investment cost and make it relatively more profitable. However, if market demand for CERs is lower than predicted the investors' profit would definitely shrink. But the good thing is that I haven't heard of any bankruptcy [of CDM renewable energy projects]. It [the financial difficulty] also depends on the type of project though. For example, small hydro-power, the investment requirement is small, even private entities do not face the danger of bankruptcy.

Interviewee 1 responded to the same question this way:

> I think it [the price drop of CERs] does not matter, because from the government's point of view, what CDM brings to China has little to do with money. Firstly, through CDM the concept of climate change is already delivered to the Chinese public, and secondly, projects that are beneficial to climate change mitigation and sustainable development are done. These are both good things.

It can be seen from the above responses that although the global financial crisis brought inconvenience to China's renewable energy CDM projects, the negative impact was not severe enough to deny the benefits CDM brought to China. This was probably why Michaelowa and Buen (2012, p. 1) described the popularity change of CDM in China as a transformation from "Cinderella to fairy princess". The "fairy princess" had certainly shown resilience even under the negative impacts of the 2008 global financial crisis and the failure of the Copenhagen conference in 2009. There was even a last minute rush for the registration of CDM projects. Due to policy change, the EU would only accept CERs generated from the least developed countries after 2012 (Michaelowa & Buen 2012, p. 30). This meant that China could only enjoy the CER benefits from pre-2012 projects. The investment in renewable energy could slow down without the stimulation of CER revenue. This was also why China started considering the feasibility of China's domestic carbon offsetting program – China CER (CCER). It was hoped that such innovation could help to sustain renewable energy investment. Apart from promoting renewable energy development, CDM also introduced a model of new market mechanism to China, which could be studied and used for China's domestic carbon market. The participation in CDM had prepared China with essential institutional, legal, implementing, monitoring, and auditing skills for the operation of ETS.

CDM as a preparation for China's national carbon market

CDM experienced ups and downs during the 11th and 12th Five-Year Plan period. As a market mechanism, it could not avoid the influence of the global economy. However, China's attitude towards CDM stayed positive even though post-2012 CDM projects mainly focused on the least developed countries. China gradually recognised the strengths of this form of project-based mechanism and adopted it as a co-program for its domestic pilot ETS. The Chinese government gave out positive political and economic signals on the tackling of climate change. Politically, it had been actively seeking policy solutions. In 2012, the State Council issued the Work Plan for Controlling Greenhouse Gas Emissions during the 12th Five-Year Plan Period (2011–2015), within which all provinces and autonomous regions were appointed individual carbon intensity (carbon emission per unit of GDP) reduction tasks. Nationally, carbon intensity was set to be reduced by 17% (18.2% achieved); new and renewable energy proportion was set to reach 11.4% of the total primary energy mix (NDRC 2012, p. 9). The central government also published the *Action Plan for Addressing Climate Change in Industry (2012–2020)*, the *National Plan for the Development of Science and Technology on Climate Change during the 12th Five-Year Plan Period*, the *Interim Measures on Low-carbon Products Certification Management*, the *Plans for Energy Development during the 12th Five-Year Plan Period*, the *Plans for the Development of Energy-Efficient and Environmental-Protection Industries during the 12th Five-Year Plan Period*, the *Suggestions on Speeding up the Development of Energy-Efficient and Environmental Protection Industries*, the *Industrial Energy Efficiency during the 12th Five-Year Plan Period*, the *2013 Implementation Plans for Industrial Energy Efficiency and Green Development*, the *Action Plan for Green Architecture*, as well as the *National Ecosystem Protection during the 12th Five-Year Plan* (NDRC 2013, p. 10). Economically, carbon trading had been decided as the desired model. Interviewee 1 described the experiment of domestic carbon market as a slow process.

> Strictly speaking, China's involvement in carbon trading started with CDM. But the domestic carbon market took many years to establish. From 2008 to 2011, the voluntary trading was not in good shape. There were no [carbon credit] buyers apart from a couple of companies that aimed at polishing company image for the Olympics. Now we have seven pilots[2] and things are getting serious. They cover different types and numbers of industries because there are such big differences between regions. This is why NDRC choose seven rather than one pilot as we aim to explore more detailed options that are suitable for China's situation.

When asked to provide insights and feedback on China's carbon trading pilot programs in late 2013, Interviewee 1 answered:

> As far as I know, only the Shenzhen cap and trade is under operation at the moment. They started on the 18th of June [2013]. The remaining six are in

the designing and planning process. But they must be in operation before the end of the year [2013] according to NDRC's deadline otherwise they don't have much time left [to compete the pilot program], only next year and the year after.[3] The core purpose of cap and trade is to give out the appropriate carbon price signal, and the appropriate carbon price signal depends upon emission reduction cost and the cap you set. The EU's experience is that the cap is too loose and the price of carbon dropped significantly. It is difficult to say what kind of cap is appropriate, as it is all according to calculations and assumptions. I've heard that the Shenzhen pilot has been very strict, but recently some entities are selling their allowance. This might indicate that they received more allowance than they needed. But we don't know how much more in total. This is really a difficult issue to handle. Therefore, we'll keep on monitoring and modifying the system.

The establishment of China's carbon trading pilot program was a positive sign which indicated that China held a more proactive attitude towards climate change mitigation. According to the insights of Interviewee 1, the Chinese government would definitely invest more towards climate change mitigation, because the direction of this investment paralleled with China's transition to a low carbon economy. The government's goal was to send out both political signals and economic signals to industries and the general public on the importance of climate change mitigation as Interviewee 1 stated below:

At the moment, the political signal is being given out, such as building ecological civilisation and transition to a low carbon economy. But the price signal is not given out yet. That's why we are having seven carbon trading pilots. Price signals such as environmental tax, carbon tax and cap and trade will give industries a clear signal to help with their decision-making on whether or not to convert to low-carbon production. We are also learning from the experience of developed countries such as the EU cap and trade and Australia's carbon tax. ...We'll certainly have a bigger move towards emission reduction and the signals will be more and more obvious.

China's carbon trading Pilots

Following the Shenzhen Pilot, the Beijing, Tianjin, Shanghai, and Guangdong Pilots had all started trading by December 2013. By mid-2014, the last two pilots, Hubei and Chongqing, formally joined the program. The implementation of all seven carbon trading pilots represented an important step for China's utilisation of market mechanisms in its green and low-carbon development. By the end of 2014, the seven pilots' total trading volume was approximately 1.26 billion yuan. Over 28.67 million tCO_2e allowance were traded in China's carbon market with per unit tCO_2e costing an average of 44.12 yuan (Zhang & Piao 2015). 2015 was an important year in the building of China's national carbon market. On the Paris Climate Change

Conference, the Chinese government announced the plan to carry out national ETS in 2017. By the end of 2015, Beijing, Shanghai, Tianjin, and Guangdong had successfully completed the second round of compliance. Hubei and Chongqing completed the first round. The total accumulated turn volume of the seven pilots exceeded 1 billion yuan. Hubei market took up the highest proportion of 42.30% trading volume, while the Shenzhen marketed accomplished 16.78%, becoming the second highest among all pilots. The Chongqing pilot achieved the least amount of trading volume, taking up only 0.22% of the total trading volume of 2015 (BJX Market Analysis 2015).

Year 2015 was also the first year for CCER, China's domestic carbon offsetting program, to enter the pilot program and participate in trading (BJX Market Analysis 2015). The seven pilots announced their own *Carbon Offsetting Management Method*, resulting in different CCER policies. The accumulated CCER trading volume of the seven pilots achieved approximately 35.48 million tCO_2e by 31 December 2015. Shanghai achieved the highest trading volume of 25.43 million tCO_2e, which took up 71.68% of total volume. Beijing listed number two in terms of CCER trading volume with 5.12 million tCO_2e and a proportion of 14.43% of the total. Chongqing was lagged behind all other pilots with no CCER trading record. The remaining four pilots took up only a small percentage of the total volume (Tanjiaoyi 2016). The CCER price also varied among the seven pilots. According to the China Beijing Environment Exchange, the CCER trading price ranged between 15 and 35 yuan. In Shanghai, the CCER trading price was lower, ranging from 12 to 20 yuan. (BJX Market Analysis 2015).

The above data shows the trading of both carbon credit and CCER varied greatly among pilots. This was partly due to different policies and approaches among different pilots. According to Interviewee 1, it was purposely set up like this to accommodate the different situations of different regions, and also to explore the pros and cons of different policies. However, the different outcomes also exposed concerns. For instance, the performance of the Chongqing pilot was not optimistic. There were only 19 transactions from 19 June 2014 to 23 July 2015. The compliance rate was only 70%. It was far below the compliance rate of the rest of the pilots. The transparency of the Chongqing pilot trading was also low. The Chongqing government only controlled the total emission volume. Participating enterprises were allowed to decide their own carbon credit allowance. However, the total volume was based on enterprises' historical highest emission record. Both of these resulted in the oversaturation of carbon credits. Trading became unnecessary. The Chongqing case raised questions for the national carbon market. Whether non-pilot regions can be ready for the 2017 national carbon market and what policy and management preparation are required deserve consideration.

Nonetheless, the overall compliance rate of China's carbon market was high in the second half of the 12th Five-Year Plan. For the Shenzhen pilot, on 30 June 2014, among the 635 enterprises which participated in the pilot carbon trading, only four failed to complete the 2013 compliance. By June 2015, only two out of the 636 participating enterprises failed to comply with the 2014 target. Those enterprises which failed to comply would receive punishment such as credit loss and removal of financial subsidies. For those who failed to payback the credit within the required time, NDRC would forcibly deduct credit from their account for the amount of their emission which exceeded the legal amount. If this was still not sufficient to payback their exceeded amount of emission, credit from the next compliance year would be deducted, and a fine three times the 2015 average carbon credit price would be applied (BJX Market Analysis 2015). The Shanghai pilot achieved 100% compliance for both 2014 and 2015. Beijing, Guangdong, Tianjin, Hubei all achieved above 95% compliance rate.

The Chinese government's consistent and systematic policy approach on climate change mitigation contributed positively to China's low-carbon development. On 19 November 2015, China's climate change special representative, Xie Zhenhua, announced the start-up of national ETS in 2017 at the State Council News Conference (Liu 2015, n.p.). This announcement corroborated the statement of Interviewee 1 in late 2013 and indicated that the Chinese government did carry out a range of experiments and investments as promised. According to the World Bank (2014, p. 38), many financial institutes such as investment banks and trading houses had started to quit the international carbon market due to the global economic downturn and reduced demand for carbon credit. Within such global environment, China continued committing to investments in renewable energy, energy efficiency improvement and other types of low carbon environment-friendly projects, which showed the strengths of consistent national climate policies and the government's determination.

Weaknesses of China's CDM and CCER program

The benefits CDM brought to China were numerous but China's CDM and CCER projects also exposed weaknesses in financial perspective and the future perspective. First, the involvement of domestic financial institutions was insufficient, and small- to medium-scale CDM and CCER projects financing appeared difficult, which created barriers for the healthy development of the national carbon market. Second, consideration needs to be paid to China's future low-cost climate change abatement options.

Insufficient involvement of domestic financial institutions

Carbon finance refers to "all financial transactions concerning reduction of carbon emissions, including bargaining transaction, investments, and speculations of carbon emission right and its derivative products" (Zou, Yang &

Yu 2011, p. 1886). In China, carbon finance is most often associated with financial activities around CDM. According to the IEA (2008, p. 224), in order to reduce global emissions by 50%–85% by 2050 from the 2005 level to avoid catastrophic climate consequences, a total of $45 trillion investment in research and development and commercialisation of a range of low carbon mitigation procedures are needed globally, and a big proportion of this investment needs to be realised in China.

The flush of carbon finance in China was brought along by foreign financial institutions such as the World Bank with the introduction of CDM (Qi 2013, p. 89). The global CER market was valued at approximately $14 billion in 2009 (UNEP, SEFI & New Energy Finance 2009, p. 16). Although the CDM-driven carbon finance was only 1%-2% of China's total energy investment, it accounted for approximate half of China's renewable energy investment (Lewis 2010, p. 2877). In this sense, international carbon finance had a significant impact in pushing forward China's renewable energy development. Domestically, the Chinese government indeed provided a large amount of funding for domestic low-carbon development in renewable energy and energy efficiency improvement; however, the majority of the funds were distributed to large-scale projects through direct or indirect investment by the central or provincial government (Qi 2013, p. 82). CDM financing for domestic small-scale enterprises appeared difficult. The smaller enterprises needed to secure loans from commercial banks as they received no direct funding from the government like the major energy state-owned enterprises (SOEs) did. But because commercial banks require 8% profitability and most small-scale CDM projects would never meet this profitability level, financial situations for China's small-scale CDM projects were not optimistic (Qi 2013, p. 90). Therefore, dealing with the financial obstacles that the small investors face has a significant meaning for the healthy development of China's small-scale carbon offset projects and the sustainability of private investment in climate-related low-carbon development.

China's Industrial Bank was the first domestic commercial bank to provide low-carbon financing to a range of energy efficiency and renewable energy projects. In 2006, China's Industrial Bank cooperated with International Finance Corporation (IFC) under the World Bank to provide funds and services to energy conservation and emission reduction projects. According to the initial agreement of the above two organisations, IFC provided 200 million yuan (approximately US$33 million) principal share of loan into Industrial Bank's 460 million yuan (approximately US$80 million) loan portfolio; Industrial Bank would only provide loans to enterprises which focused on energy efficiency, renewable energy, and emission reduction in general or those approved by IFC (Zou, Yang & Yu 2011, p. 1887). By September 2008, IFC had provided 2.8 billion yuan (approximately US$470 million) in total to the Industrial Bank to finance China's energy saving and emission reduction projects, which was predicted to reduce over 10 million tons of carbon emission (Liu & Shen 2011, p. 215).

The involvement of IFC was an important causal factor to the Industrial Bank's initiative in carbon finance. IFC signed a loan loss sharing agreement with the Industrial Bank. It worked as an effective risk compensation mechanism which reduced risks for the Industrial banks (Liu & Shen 2011, p. 215). For instance, the Industrial Bank injected 1.5 billion yuan (approximately US$230 million) to low-carbon financing during the second phase of the joint agreement with IFC during 2007–2008. IFC provided US$100 million as risk-sharing fund during the same period (Qi 2013, p. 88). However, without the involvement of international finance institutions, it was doubtful if a domestic commercial bank would independently invest large amounts of funding into carbon finance due to numerous risks associated with the carbon market. By the end of the 12th Five-Year Plan, only a few Chinese commercial banks dealt with CDM and CCER loans. Industrial Bank is the first one which carried out detailed research into CDM finance. According to the product development manager, Cao Bin, Industrial Bank was the first commercial bank to launch a carbon financing product kit after their comprehensive study on CDM market since 2008, which aimed at providing diversified banking services to key CDM stakeholders (CDM China 2010, n.p.).

The Industrial Bank's move was a significant step forward as a domestic commercial bank to specialise in loans for low-carbon enterprises. However, the amount of funds available was still insufficient compared to the vast demand from the low-carbon sector such as the renewable energy and energy efficiency projects (Liu & Shen 2011, p. 214). Following the Industrial Bank's initiative, a couple of small banks followed suit. For instance, Shanghai Pudong Development Bank launched Green Credit Service to enterprises in the low-carbon development category. In 2009, it independently provided CDM-related consultation and services to two hydroelectric projects in Shanxi Province. Investment Banking, a subbranch of the Agricultural Bank, provided financial assistance and services to 10 CDM projects across the country (Zou, Yang & Yu 2011, p. 1887). On 11 December 2014, Bank of Shanghai, partnered with Shanghai Environment and Energy Exchange, signed the first CCER pledge loan agreement of 5 million yuan with Shanghai Baocarbon New Energy and Environmental Protection Technology Co., Ltd. Bank of Shanghai also promised 3 billion yuan carbon finance to the clients of Shanghai Environment and Energy Exchange over the following three years, which looked more promising than previous inputs.

In theory, carbon finance should play an important role in CDM and CCER and the development of a healthy national carbon market (Amran, Zainuddin & Zailani 2013; Labatt & White 2007); however, the participation of China's domestic financial institutions was very limited. Fuhr and Lederer (2009, p. 337) noticed that Chinese banks "have no interest in [carbon] trading, hardly an interest in project financing". This statement described the attitude of majority Chinese banks; however, it did not dig deep to uncover the fundamental reasons behind the Chinese banks' stance. The fact was it was not that the Chinese banks were unwilling to get involved in carbon finance, but they did not know how to do it and feared the risks of doing so.

Carbon finance was a relatively new phenomenon to Chinese banks. It was only introduced to China roughly ten years ago. Chinese banks were extremely cautious with carbon finance due to the lack of understanding of CDM, CCER, and the associated market and policy risks. Lack of expertise and experience was the first practical barrier for Chinese commercial banks to get involved in carbon finance. As Heggelund (2007, p. 185) put it, "lack of market understanding and awareness among decision makers and stakeholders in the financial sector could be one factor affecting China's allocation of projects". The lack of understanding had made Chinese financial institutions reluctant in developing financial products and providing financial services related to emission reduction and climate change (Shen 2015, p. 347). According to Lo and Yu's (2015, p. 375) interview with a Chinese commercial banker, it was not that Chinese commercial banks had no interest in the carbon market. They did not embrace carbon finance because they "don't understand the market", "don't know how much could be lost", and "can't make judgments about market trends". Moreover, there were usually multiple parties involved in CDM projects, which further complicated the relationship among domestic and international financial institutions involved. It might even involve the application of law and territorial jurisdiction if disputes occur over the contracts and implementation processes (Zou, Yang & Yu 2011, p. 1889). However, CCER projects reduced the complexity compared with CDM because all parties operated within the national carbon market. This was a more promising aspect for Chinese domestic banks with regards to carbon finance.

Second, the registration procedure for CDM projects was strict and complicated, which was to say, a project applying for CDM might not be necessarily granted an approval if it failed to demonstrate any CDM registration criteria. In fact, the EB approval rate of Chinese CDM projects was quite low (Wang & Chen 2010, p. 1994). This made Chinese commercial banks to further consider the risk factor of providing loans to some types of CDM projects, such as solar and wind energy projects. This type of renewable energy project might only be borderline profitable with the sale of CERs; therefore, commercial banks would had to consider the consequences if they failed to register with CDM.

Third, market and policy risk was external influencer for Chinese commercial banks' decision making. After the 2008 global financial crisis, many international investment banks and trading houses began to exit the global carbon market, and project developers also reduced investment in carbon abatement due to the declined demand of carbon credits created under Kyoto Protocol (The World Bank, 2014, p. 38). The uncertainty over the future of CDM under the Paris Agreement further intensified this problem (Hone 2017, n.p.). The carbon price, as a consequence, dropped dramatically. The fluctuation of the carbon price and the majority of big global investment banks' behaviour both sent negative signals to Chinese commercial banks, and therefore, further influenced their decision in carbon finance.

Last, the inconsistency of international climate policy added uncertainty to future carbon market. After 2012, the market demand for CER largely shrank. CERs were only purchased if a country decided to include foreign CERs in its domestic emission reduction policy or CDM continued as a part of post-2012 Kyoto Protocol (Lewis 2010, p. 2876). Although a number of domestic cap and trade programs allowed international carbon offset, the allowance of percentage of CERs was restricted. For instance, the EU only allowed emitting enterprises to purchase CERs from projects approved before 2012 if absence of post Kyoto agreement, and these CERs must be from the projects types approved by all member states and could only be used through to 2014 (European Commission 2008, n.p.). Regional trading scheme such as California's Assembly Bill 32 also restricted the criteria for the purchase of future CERs (RGGI 2009, n.p.). The Paris Agreement did not identify whether or not CDM would be used as an ongoing international offsetting mechanism (Hone 2017, n.p.). As a result, the international outlook for future CERs market was not optimistic, which also contributed to Chinese commercial banks' reluctance in the investment of carbon finance. However, the market and policy risks associated with CDM can be reduced if the program is downgraded from international offset scheme to a well-regulated national scheme, given that the Chinese government would provide consistent and supportive national policies. In Chapter 6, this book will project the financial outlook of CDM's offspring, China's domestic carbon offsetting program CCER, and will explain what a supportive financial system means to the CCER program and China's national ETS.

Impact of financial difficulty on China's CDM renewable energy projects

The passive attitudes of China's domestic financial institutions created financial difficulties for many of China's small-scale low-carbon development, especially in that of the renewable energy projects. Most renewable energy projects had higher upfront investment requirements compared with coal-fired power plants, and therefore took longer to gain profit (Lewis 2009, p. 1208). As a result, they were more vulnerable without long-term financial support. According to Interviewee 6, only at the start of China's CDM program, some foreign buyers would provide upfront finance for establishing the projects (bilateral projects). For many of the later projects (unilateral projects), especially the small ones, project developers would have to source loans independently:

> This appeared difficult for them as they could only obtain CER revenue after production, well, if they could find buyers. How much they can sell their CERs for was also depending on the market situation. Banks had to consider these risks when assessing the project developers' repayment ability.

Some CDM-related renewable energy manufacturing enterprises also faced the same financial situation. The majority of domestic commercial banks had stopped lending to PV solar manufacturing industry due to the decreasing demand for PV solar equipment in the last couple of years. Only a small number of domestic commercial banks continued to provide loans; however, the maximum loan term was only five years (Huang 2014, n.p.). Some companies had to ask local governments to act as a middle man to negotiate with banks for loans. For instance, Shenzhen Topray Solar was one of the top PV solar panel manufacturers in China. The company's representative expressed that lack of financing was the biggest obstacle restricting the company's development. In 2014, Topray had no choice but to submit a petition to the Shenzhen City government wishing the government would communicate with the National Development Bank and Import and Export Bank and possibly used administrative power to persuade them to lend (Huang 2014, n.p.). The financial difficulty was a common phenomenon for many non-state-owned Chinese renewable energy enterprises and related manufacturing enterprises. Take the low-carbon investment during the 11th Five-Year Plan (2006–2010) for example, it required 800 billion yuan (approximately US$124 billion) low-carbon investment annually to achieve the 20% energy intensity reduction. However, the Chinese central budget only allowed 83.3 billion yuan (approximately US$13 billion) annually for low-carbon activities, and the majority of these funds were distributed to large SOE projects. This was to say, (1) the government budget alone was insufficient to meet the requirement; (2) small non-state-owned low-carbon projects were almost excluded from direct state financial support and had to seek financial products elsewhere. International financial organisations, some commercial banks, and private entities had indeed provided large proportion of loans to China's low-carbon projects during this period; however, they were in more favour of large-scale projects with matured technology and operation as a way of lowering risk. Small-scale projects proposed by private entities such as renewable energy projects (including CDM projects), therefore, faced a difficult financial situation – limited access to both state and non-state funding. Lack of financial assistance for private small-scale low-carbon projects was a significant flaw in China's renewable energy policy and overall low-carbon development strategy. This was because, without finance, China's renewable energy development was unlikely to be sustained. The government had realised the important role of private investment in renewable energy industry as stated in the *National Energy Administration No. 179 Document* (NEA 2012), thus financial availability for small investors should be made the priority discussion in future policy considerations.

Future low-cost abatement potential

CDM introduced a great concept of pursuing climate change mitigation and sustainable development at the same pace. Its positive impact on Chinese society was obvious. There might be some concerns that CDM exhausted

China's cheap climate change abatement opportunities, but compared with the overall benefit CDM brought to China, it was very minor and could be overcome.

According to Castro (2012, p. 212), China had picked 32% of its low-hanging fruit through CDM, which was a very high proportion compared with other CDM host countries. This might potentially put pressure on China's future emission abatement opportunities. However, the 32% was only of the identified theoretical abatement potential for CDM projects. Looking beyond the pre-identified categories (industrial gas, new and renewable energy, energy efficiency, and fuel substitute projects) there were still plentiful low-cost abatement opportunities. According to Jiang and Tovey (2010), the building sector, which accounted for 25% of China's total emissions, left China plenty of opportunities for future emission abatement, as the building sector was almost untouched by CDM. Other abatement options such as contracted energy consumption reduction or efficiency improvement for households and companies, localised land-use change, and small-scale revegetation based on family and community units in rural areas could also open up new opportunities for China's future emission reduction activities. Even within the renewable energy sector, abatement opportunities still exist and may upscale as long as the demand for CERs/CCERs exists and their price high enough to cover the gap between renewable energy and fossil fuels. As technologies advance, the cost of abatement will drop simultaneously and more low-cost abatement will emerge (Michaelowa & Buen 2012, p. 26).

Moreover, abatement cost reduction can also be achieved from simplified administrative measures such as reduction in transaction costs. The high transaction costs associated with CDM projects were often considered a weakness of CDM. The transaction cost of CDM was the total expenditure on completing a transaction of CERs between buyers and sellers. It included project search cost, project document development cost, negotiation cost, validation cost, registration cost, monitoring cost, verification and certification cost, and share of proceeds (The World Bank et al. 2004, p. xxxii). Due to the existence of the high transaction cost, the majority of Chinese CDM projects, and especially the ones that were popular with CER buyers, were often big in scale with high CER returns. The transaction cost imposed great difficulty for many small-scale projects and made them uneconomical and unlikely to register with CDM. According to the analysis of Michaelowa and Jotzo (2005) on early CDM projects, small projects with annual CER production under 10,000 would have difficulty to cover the transaction cost from the sales of CERs. In this sense, many small-scale projects were left out by the current CDM program and conserved for the future. For example, Interviewee 7 from the Centre for Rural Environmental and Social Research, Chinese Academy of Social Sciences (CASS) mentioned that partial reforestation and the return to traditional pattern of land use in certain rural regions of China would bring great benefit to local environment and at the same time beneficial to carbon emission reduction.

> Yunnan [province] has been dry for quite a while. How can this place be dry? This is the last place nationwide which should be dry. This region undoubtedly has a very complex ecology. According to our research, we found that everywhere else is dry apart from the traditional terraced fields with forests on top of the mountains. This is because the forest on top of the mountains has a great ability in conserving water resources. But in many cases, vegetation has been completely cleared. So we are now hoping to relieve the drought through converting farmland on mountain tops into forests. The forests will not only absorb water and hold on the soil but can also absorb CO_2e and contribute to carbon emission reduction, so I think this will be a very good project for the local communities.

The reforestation project mentioned by the above interviewee is a typical example of small-scale abatement options that can bring solid benefit to local communities. But according to NDRC statistics, among all approved CDM projects, there were only five projects within the afforestation and reforestation category, leaving vast potential for the future (NDRC 2014). This type of small-scale reforestation projects can be delivered through the unilateral type of CCER projects. Unilateral CCER projects involve solely the host party throughout the planning and implementation process, and therefore impose a lower transaction cost (Maraseni & Gao 2011, p. 340; Michaelowa 2007). If further government support is provided, such as giving priority to CCERs of small-scale projects to be traded on the domestic carbon trading pilot program, more low-cost community-based abatement opportunities will be unearthed. Therefore, CDM has not drained China's low-cost abatement opportunities. If financial institutions, services, and the trading system mature, such small-scale abatement projects can be absorbed into the national carbon market. There are also similar types of potential projects in the building and transport sector that deserve further investigation and could be utilised in China's national carbon market.

Summary and future outlook

This chapter investigated what CDM meant for China's climate change mitigation and sustainable development. This investigation was of significance because CDM was the first large-scale international climate change mitigation mechanism China participated in. It not only opened a door for advanced technologies, project design methodologies, and monitoring and auditing procedures, but also brought in the concept of carbon market and how to utilise carbon market to achieve climate change mitigation and sustainable development. Therefore, the study of the impacts of CDM on China's society was valuable. Overall, CDM was more beneficial than detrimental to China's low-carbon development goals. This chapter focused on two types of energy-related CDM projects: renewable energy projects and energy saving and emission reduction projects. Both targeted the long-lasting greenhouse gas CO_2. Although they did not directly contribute to China's emission reductions, they brought China a range of practical benefits. First of all, CDM helped to raise business awareness of climate change among corporates which participated in CDM, which echoed the trends of climate awareness improvement among the public. Second, CDM stimulated China's renewable energy development, which had a positive effect on the optimisation of China's energy structure. The last important benefit of CDM was that it introduced the concept of the carbon market, and it put this concept into practice. Through this, CDM paved the way to China's carbon trading pilot program and national ETS. According to insights of the NDRC official whom this book interviewed, CDM-like carbon offsetting program would not end in China by 2012. During the writing of this book, the CCERs were indeed incorporated into China's carbon trading pilot program and national ETS. Although CDM was not granted a definite future in the Paris Agreement, China was not impacted by this international decision. At the symposium, *The Future of Clean Development Mechanism*, jointly held by the China Carbon Forum and Sindicatum Sustainable Resources, the Chinese representatives[4] confirmed the role of CDM in the future. When asked "Was there even a future for CDM?", the Chinese representatives stated:

> From a business point of view, it is difficult to change the fundamentals of oversupply. On the other hand, the development of carbon markets is flourishing, so the opportunity for CER uptake can only grow. (China Carbon Forum 2013)

It is certain that CCERs has an important role in China's national ETS which allows industries to identify low-cost abatement opportunities domestically and use CCERs for their compliance. There is also the possibility for CERs to be absorbed in China's national carbon market (China Carbon Forum 2013).

Nonetheless, two issues associated with Chinese CDM were exposed. One was the financial difficulty that small-scale renewable energy CDM projects experienced. The other was the concern that CDM exhausted China's low-cost abatement opportunities. Both issues were linked to the nature of CDM to some extent. CDM, as an international mechanism, required high transaction costs; therefore, large-scale projects appeared more economical. As a result, the Chinese government and financial institutions purposely gave priority to large-scale projects with lower risks. However, many small-scale projects, especially those in renewable energy, were in fact more effective for sustainable development and brought more practical benefits to the local community and environment. Even the Chinese government admitted that small private investors in renewable energy were a crucial element of China's future energy market and carbon market. Therefore, small-scale low-carbon investors' financial difficulty must be treated as an urgent issue and be dealt with in a timely manner.

The second concern over the exhaustion of China's low-carbon abatement opportunity was less crucial compared with the first one. Again, abolishing the "big-scale obsession" is the key. It is still possible to discover abundant and diversified small-scale mitigation opportunities. To include and promote sustainability enhancing small projects, the government could incorporate a range of flexible measures and policies to simplify project approval procedures and reduce transaction fees. It is certain that the policy and market risks can be reduced if CDM is downgraded to CCER as a co-mechanism of China's national ETS. In sum, the use of CDM as a policy response to climate change mitigation in China left significant scope for discussion and future exploration of the issues highlighted in this chapter. In the next chapter, this book will focus on a parallel domestic climate-related policy, the Energy Conservation and Emission Reduction Scheme (ECERS). Its strengths and weaknesses will be explored in detail.

List of Interviewees

Interviewee 1, Officer of NDRC

Interviewee 6, ENGO officer Quanlian New Energy Chamber of Commerce

Interviewee 7, Professor at the Centre for Rural Environmental and Social Research, CASS

Notes

1. This refers to a series of standards developed and published by the International Organisation for Standardisation (ISO), which define, establish, and maintain an effective quality assurance system for manufacturing and service industries.
2. The seven cap and trade pilot programs in China are Beijing, Tianjin, Shanghai, Chongqing, Shenzhen, Guangdong province and Hubei province.
3. Up till April 2015, seven carbon trading pilots have all been in operation with an accumulated total turnover 1.3 billion yuan. NDRC plans to start a national trading scheme in the end of 2017.
4. Key representatives from the Ministry of Foreign Affairs, China CDM Fund, Sino-Carbon Innovation & Investment and Enecore joined the panel discussion. Sixty participants from China's climate change community were part of the audience.

References

Amara, N, Ouimet, M & Landry, R 2004, 'New evidence on instrumental, conceptual, and symbolic utilization of university research in government agencies', *Science Communication*, vol. 26, no. 1, pp. 75–106.

Amran, A, Zainuddin, Z, Zailani, SHM 2013, 'Carbon trading in Malaysia: review of policies and practices', *Sustainable Development*, vol. 21, pp. 183–192.

Belis, D 2013, 'The role and dynamics of the Clean Development Mechanism in EU-Vietnam climate relations', in H Bruynincks & D Belis (eds), *The governance of climate change relationships between Europe and Asia*, Edward Elgar Publishing, Cheltenham, pp. 157–188.

BJX Market Analysis 2015, *2015 zhongguo tanshichang fenxi (2015 China carbon market analysis)*, viewed 23 November 2018, <http://mhuanbao.bjx.com.cn/mnews/20160826/766956.shtml>.

Boyd, E, Hultman, N, Roberts, JT, Corbera, E, Cole, J, Bozmoski, A, Ebeling, J, Tippman, R, Mann, P, Brown, K & Liverman, DM 2009, 'Reforming the CDM for sustainable development: lessons learned and policy futures', *Environmental Science & Policy*, vol. 12, no. 7, pp. 820–831.

Buen, J & Castro, P 2012, 'How Brazil and China have financed industry development and energy security initiatives that support mitigation objectives', in A Michaelowa (ed), *Carbon markets or climate finance?*, Routledge, Abingdon, pp. 53–91.

Castro, P 2012, 'Does the CDM discourage emission reduction targets in advanced developing countries?', *Climate Policy*, vol. 12, no. 2, pp. 198–218.

Castro, P, Hayashi, D, Kristiansen, KO, Michaelowa, A & Stadelmann, M 2011, 'Scoping study: linking RE promotion policies with international carbon trade', International Energy Agency – Renewable Energy Technology Deployment, viewed 10 February 2014, <https://www.osti.gov/etdeweb/servlets/purl/21507911>.

CDM China 2010, 'Carbon trading financing myth waits to be explained', viewed 22 May 2012, <http://cdm.ccchina.gov.cn/web/NewsInfo.asp?Newsld=4478>.

China Carbon Forum 2013, The future of the Clean Development Mechanism, viewed 2 march 2017, <http://www.chinacarbon.info/392/>.

Chinese Wind Energy Association (CWEA) 2010, Statistics of China's wind power installed capacity, CWEA, Beijing.

Department of Climate Change, NDRC 2015, Clean Development Mechanism in China, viewed 26 June 2015, <http://cdm-en.ccchina.gov.cn/list.aspx?clmId=38>.

Dutschke, M & Michaelowa, A 1998, *Creation and sharing of credits through the Clean Development Mechanism under the Kyoto Protocol*, HWWA Discussion Paper 62, Institut Für Wirtschaftsforschung, Hamburg.

European Commission 2008, Decision of the European Parliament and of the Council on the effort of Member States to reduce their greenhouse gas emissions to meet the Community's greenhouse gas emission reduction commitments up to 2020, viewed 18 November 2015, <http://eur-lex.europa.eu/LexUriServ/LexUriServ>.

Evans, S 2014, 'IEA: Four charts that show what a solar powered future looks like', *Carbon Brief*, viewed 25 November 2015, <http://www.carbonbrief.org/iea-four-charts-that-show-what-a-solar-powered-future-looks-like>.

Fuhr, H & Lederer, M 2009, 'Varieties of carbon governance in newly industrialising countries', *The Journal of Environment & Development*, vol. 18, pp. 327–345.

Global Wind Energy Council 2011, *Global wind report*, viewed 10 February 2014, <http://gwec.net/wp-content/uploads/2012/06/GWEC_annual_market_update_2010_-_2nd_edition_April_2011.pdf>.

Greiner, S 2009, 'CDM: from policy to practice', *Climate Policy*, vol. 9, no. 6, pp. 681–683.

Gtowacki Law Firm 2013, CERs and ERUs market as from 2013, viewed 14 January 2014, <http://www.emissions-euets.com/cers-erus-market-as-from-2013>.

Haites, E, Duan, M & Seres, S 2006, 'Technology transfer by CDM projects', *Climate Policy*, vol. 6, no. 3, pp. 327–344.

Heggelund, G 2007, 'China's climate change policy: domestic and international developments', *Asian Perspective*, vol. 31, no. 2, 2007, pp. 155–191.

Hone, D 2017, Saving the CDM, viewed 1 October 2017, <https://blogs.shell.com/2017/05/29/saving-the-cdm/>.

Huang, S 2014, *Xinnengyuan qiye zaoyu qianhuang (New energy enterprises experience financial difficulties)*, viewed 20 November 2015, <http://energy.people.com.cn/n/2014/0705/c71661-25242441.html>.

IEA 2008, *Energy technology perspectives 2008*, IEA/OECD, Paris.

IEA 2014, *Technology roadmap: solar photovoltaic energy*, OECD/IEA, Paris.

Jiang, P & Tovey, K 2010, 'Overcoming barriers to implementation of carbon reduction strategies in large commercial buildings in China', *Building and Environment*, vol. 45, pp. 856–864.

Kang, J, Yuan, J, Hu, Z & Xu, Y 2012, 'Review on wind power development and relevant policies in China during the 11th Five-Year-Plan period', *Renewable and Sustainable Energy Reviews*, vol. 16, no. 4, pp. 1907–1915.

Labatt, S & White, RR 2007, *Carbon finance: the financial implications of climate change*, John Wiley & Sons, Hoboken, NJ.

Lewis, JI 2009, 'Climate change and security: examining China's challenges in a warming world', *International Affairs*, vol. 85, no. 6, pp. 1195–1213.

Lewis, JI 2010, 'The evolving role of carbon finance in promoting renewable energy development in China', *Energy Policy*, vol. 38, no. 6, pp. 2875–2886.

Lin, BQ & Liu, JH 2010, 'Estimating coal production peak and trends of coal imports in China', *Energy Policy*, vol. 38, no. 1, pp. 512–518.

Lin, H & Gil, J 2016, 'Confucian Principles to Enhance China's Environmental Governance', in C Andressen (ed), *Changing economies: the interplay between global and Chinese economic change and their impact on Australia*, Routledge, Oxford, pp. 136-157.

Liu, JG & Shen, ZQ 2011, 'Low carbon finance: present situation and future development in China', *Energy Procedia*, vol. 5, pp. 214–218.

Liu, Y 2015, *Jie Zhenhua: quanguo tanshichang yuji 2017 nian qidong (Jie Zhenhua*: Nationwide carbon market is planned to kick start in 2017), viewed 24 November 2017, <http://news.xinhuanet.com/local/2015-11/20/c_128448793.htm>.

Lloyd, B & Subbarao, S 2009, 'Development challenges under the Clean Development Mechanism (CDM)—Can renewable energy initiatives be put in place before peak oil?', *Energy Policy*, vol. 37, no. 1, pp. 237–245.

Lo, AY & Yu, X 2015, Climate for business: opportunities for financial institutions and sustainable development in the Chinese carbon market, *Sustainable Development*, vol. 23, no. 6, pp. 369–380.

Maraseni, TN & Gao XQ 2011, 'An analysis of Chinese perceptions on unilateral Clean Development Mechanism (uCDM) projects', *Environmental Science & Policy*, no. 14, pp. 339–346.

Michaelowa, A 2007, Unilateral CDM – can developing countries finance generation of greenhouse gas emission credits on their own? *International Environmental Agreements*, vol. 7, pp. 17–34.

Michaelowa, A 2012, 'Strengths and weaknesses of the CDM in comparison with new and emerging market mechanisms', *Paper No. 2 for the CDM Policy Dialogue*, viewed 18 March 2014, <http://www.perspectives.cc/typo3home/groups/15/Publications/2012/2012_Strengths-and-weaknesses-of-the-CDM-in-comparison-with-new-and-emerging-market-mechanisms.pdf>.

Michaelowa, A & Buen, J 2012, 'The Clean Development Mechanism gold rush', in A Michaelowa (ed), *Carbon markets or climate finance?*, Routledge, Abingdon, pp. 1–38.

Michaelowa, A & Jotzo, F 2005, 'Transaction costs, institutional rigidities and the size of the clean development mechanism', *Energy Policy*, vol. 33, pp. 511–523.

NDRC 2004a, *Interim Measures for the Provisional Rules for Management of CDM Projects in China*, viewed 19 November 2015, <http://cdm.ccchina.gov.cn>.

NDRC 2012, China's Policies and Actions for Addressing Climate Change (2012), NDRC, Beijing.

NDRC 2013, China's Policies and Actions for Addressing Climate Change (2013), NDRC, Beijing.

NDRC 2014, *China's Policies and Actions on Climate Change 2014*, viewed 03 July 2015, <http://en.ccchina.gov.cn/archiver/ccchinaen/UpFile/Files/Default/20141126133727751798.pdf>.

NDRC 2015, Expected average annual CERs by scope, viewed 21 October 2015, <http://cdm-en.ccchina.gov.cn/NewItemTable8.aspx>.

NDRC 2015a, Approved projects by scope, viewed 21 October 2015, <http://cdm-en.ccchina.gov.cn/NewItemTable9.aspx>.

NEA 2012, *Guojia nengyuanju guanyu gouli he yindao minjian ziben jinyibu kuoda nengyuan lingyu touzide shishiyijian, [Guojia Nengyuanju 179 hao] (National Energy Administration's recommendations to encourage and guide private investment in the energy sector, [National Energy Administration No. 179 Document])*, viewed 2 August 2015, <http://www.nea.gov.cn/2012-06/20/c_131665600.htm>.

Nussbaumer, P 2009, 'On the contribution of labelled Certified Emission Reductions to sustainable development: a multi-criteria evaluation of CDM projects', *Energy Policy*, vol. 37, no. 1, pp. 91–101.

Olsen, KH & Fenhann, J 2008, 'Sustainable development benefits of clean development mechanism projects: a new methodology for sustainability assessment based on text analysis of the project design documents submitted for validation', *Energy Policy*, vol. 36, no. 8, pp. 2819–2830.

Pearson, P 2007, 'Market failure: why the Clean Development Mechanism won't promote clean development', *Journal of Cleaner Production*, vol. 15, pp. 247–252.

Qi, Y 2013, *Annual review of low-carbon development in China: 2010*, World Scientific Publishing, Singapore.

RGGI 2009, The website of the Regional Greenhouse Gas Initiative, viewed 18 November 2015, <http://www.rggi.org/offsetsS>.

Schneider, L 2007, 'Is the CDM fulfilling its environmental and sustainable development objectives? An evaluation of the CDM and options for improvement', *Öko-Institut for Applied Ecology, Berlin*, vol. 248, pp. 1685–1703.

Schneider, M, Holzer, A, & Hoffmann, VH 2008, Understanding the CDM's contribution to technology transfer, *Energy policy*, vol. 36, no. 8, pp. 2930–2938.

Schneider, L 2009, 'Assessing the additionality of CDM project: practical experience and lessons learned', *Climate Policy*, vol. 9, no. 3, pp. 242–254.

Schröder, M 2012, *Local climate governance in China: hybrid actors and market mechanisms*, Palgrave MacMillan, Hampshire.

Scotney, R, Chapman, S, Hepburn, C & Jie, C 2012, *Carbon markets and climate policy in China: China's pursuit of a clean energy future*, Climate Bridge, viewed 22 May 2014, <http://www.climateinstitute.org.au/verve/_resources/ClimateBridge_CarbonMarketsandClimatePolicyinChina_October2012.pdf>.

Schwank, O 2004, *Concerns about CDM projects based on decomposition of HFC-23 emissions from 22 HCFC production sites*, Infras, Zurich.

Seres, S, Haites, E & Murphy, K 2009, 'Analysis of technology transfer in CDM projects: an update', *Energy Policy*, vol. 37, pp. 4919–4926.

Shen, W 2015, 'Chinese business at the dawn of its domestic emissions trading scheme: incentives and barriers to participation in carbon trading', *Climate Policy*, vol. 15, no. 3, pp. 339–354.

State Council of PRC 2012, *Jieneng jianpai "shierwu" guihua – guofa [2012] 40 hao (The plan for energy saving and emission reduction of the 12th Five-Year Plan, [2012 State Council Office Document no. 40])*, viewed 21 October 2015, <http://www.gov.cn/gongbao/content/2012/content_2217291.htm>.

Sutter, C & Parreño, JC 2007, 'Does the current Clean Development Mechanism (CDM) deliver its sustainable development claim? An analysis of officially registered CDM projects', *Climatic change*, vol. 84, no. 1, pp. 75–90.

Tanjiaoyi 2016, *2015 nian quanguo tanjiaoyi shidian shichang CCER yunxing qingkuang fenxi (The analysis of 2015 implementation of national carbon pilots)*, viewed 28 November 2018, <www.Tanjiaoyi.com/article-17711-1.html>.

The National People's Congress of PRC 2013, *Zhonghua renmin gongheguo kezaisheng nengyuan fa (The renewable energy law of PRC),* viewed 7 July 2017, <http://www.npc.gov.cn/npc/zt/qt/2015zhhbsjx/2013-12/04/content_1935957.htm>.

The World Bank, Ministry of Science and Technology, P.R. China, The Deutsche Gesellschaft für Technische Zusammenarbeit, German Technical Cooperation Unit (GTZ), Federal Ministtry of Economic Cooperation & Development Swiss State Secretariat for Economic Affairs 2004, *Clean Development Mechanism in China: taking a proactive and sustainable approach*, 2nd edn, The World Bank, Washington, DC.

The World Bank 2014, *State and trends of carbon pricing 2014*, The World Bank Group, Washington, DC.

UNEP, SEFI & New Energy Finance 2009, *Global trends in sustainable energy investment*, viewed 4 November 2015, <http://www.unep.org/pdf/Global_trends_report_2009.pdfS>.

Wang, Q & Chen, Y 2010, 'Barriers and opportunities of using the clean development mechanism to advance renewable energy development in China', *Renewable and Sustainable Energy Reviews*, vol. 14, pp. 1989–1998.

Xuan, XW & Xue, JJ 2012, 'China's green low-carbon development', in JJ Xue, ZX Zhao, YD Dai & B Wang (eds), *Annual report on China's low carbon economic development 2012*, Social Sciences Academic Press, Beijing, pp. 1–29.

Yin, ZQ, Wang, SC, Zhao, Y, Wang, ZF, Zheng, RC, Zhong, JS, Hu, RQ, Zhou, XW, Li, ZC, Zhu, CH & Gao, G 2012, 'PV solar energy', in F Li & Z Zhang (eds), *China's medium to long term (2030, 2050) energy development strategy research*, Science Publishing, Beijing, pp. 165–250.

Zhang, X & Piao, LJ 2015, 'Quangguo qi shengshi tanjiaoyi shidian yanjiu (Research of the seven carbon trading pilots nation-wide)', *Zhongguo Chuanjian*, vol. 1, viewed 23 November 2018, <http://wap.cnki.net/qikan-ZGCJ201501036.html>.

Zou, F, Yang, H & Yu, LQ 2011, 'Thinking on the development of "Carbon Finance" in commercial banks of China', *Energy Procedia*, vol. 5, pp. 1885–1892.

5 Energy conservation and emission reduction scheme (ECERS) for the energy production sector and energy users

Introduction

The Chinese government implemented the Energy Conservation and Emission Reduction Scheme (ECERS) from the beginning of the 11th Five-Year Plan for National Economy and Social Development (2006–2010) to relieve domestic energy pressure and as a response to the international calls on climate change mitigation. By the end of 2015, ECERS had been implemented for ten years (2006–2015). It continued to be carried out through the current Five-Year Plan, the 13th Five-Year Plan. ECERS focused on energy saving and emission reduction through all sectors, which included the energy sector, the manufacturing sector, the transportation sector, the building sector, and the agriculture and forestry sector (The Central People's Government of PRC 2006a, n.p.). Among all these sectors, the energy production sector accounted for one of the largest proportions of China's CO_2 emission. For instance, in 2008, the energy sector alone consumed 920 million tons of coal equivalent (Mtce) and generated 2.5 giga tons (Gt; 1 Gt = 1 billion tons) of CO_2 emission, which took up 42% of national energy-related CO_2 emission (Qi 2013, p. 105). Also, energy was consumed by other energy-intensive sectors such as the manufacturing sector and the building sector. Thus, the carbon-intensive nature of the energy sector was the fundamental reason for China's overall high emissions. Due to this status, the energy production sector deserves special attention and is treated as a focal point of this chapter. China's current energy structure is still dominated by coal; therefore, utilising coal more efficiently in the process of energy production would produce positive impacts on resource conservation and emission reduction. Simultaneously, due to this coal-dominated energy structure, how end users utilise energy also matters. Therefore, a newly initiated incentive-based program, the Energy Performance Contracting (EPC) program, targeting energy end users, will also be discussed. It is anticipated that such investigations can build up the answer to the question of how China can achieve better climate change mitigation through improving its energy policies.

The goal of this chapter is to analyse the strengths and weaknesses of ECERS for energy producers and major energy consumers during the scheme's implementation from 2006 to 2015 covering China's 11th and 12th Five-Year Plans. This time frame roughly echoes that of the Clean Development Mechanism (CDM), the parallel international climate change mitigation scheme that China participated in, as mentioned in Chapter 4. The energy producers refer to China's predominant thermal power producers and new and renewable energy producers. Energy consumers refer to enterprises which consume electricity power. Within the energy production sector, ECERS emphasised efficiency improvement and the optimisation of energy structure, which included replacing inefficient small thermal plants with large-scale more efficient ones and promoting the development of renewable energies. For energy consumers, ECERS focused on each entity's efficiency improvement in utilising energy such as the installation of energy saving equipment in factories and sustainable energy management for businesses, schools, and office buildings.

ECERS in the energy production sector

ECERS in the energy generation sector was the most important component of the scheme because it tackled climate change from the deep roots of energy production. It aimed at improving the efficiency of conventional coal-fired power plants and increasing the proportion of new and renewable energies (The Central People's Government of PRC 2006a, n.p.). Due to the fact that approximately 70% of China's energy was generated from coal (Li et al. 2011, p. 568), efficiency improvement in the coal-power sector was significant to China's CO_2 emission mitigation. The increasing promotion of renewable energies optimised the energy structure and benefited long-term sustainability.

During the 11th Five-Year Plan (2006–2010), ECERS achieved overall positive results in the energy production sector. Approximately 768.3 million kW capacity of inefficient small thermal plants were shut down nationwide. The average coal consumption per each kilowatt hour (kWh) of electricity dropped from 370 g to 333 g (China Electricity Council 2011, n.p.). The proportion of combined non-fossil energy capacity achieved 26.6% of the national total, which increased by 1.6% compared with that at the end of the 10th Five-Year Plan (2000–2005). The increase of non-fossil energy proportion in the nation energy structure accumulatively replaced 1.5 billion tons of coal and reduced approximately 3 billion tons of CO_2 emission. Wind power and solar power were two of the fastest developed non-fossil energies with wind power capacity reaching 310 million kW and solar power capacity reaching 6 million kW (State Grid Corporation of China 2012, n.p.). The 12th Five-Year Plan continued to work on these areas. At the end of the 12th Five-Year Plan, in 2015, to be exact, the Chinese government announced a plan to restrain coal production and coal-fired power production, which further reduced the share of coal in China's energy structure.

This looked optimistic, but are there problems and deficiencies with the scheme? What has caused them? What are the impacts of the problems? An investigation of the scheme's strengths and weaknesses in the energy production sector is very necessary. This is because energy production is one of the most critical factors that impacts on the social economy and people's livelihood; and more pressingly, energy structure, as a fundamental indicator for a country's CO_2 emission, its optimisation can bring a very positive outcome to climate change mitigation. This is especially applicable for China due to the scale of its energy demand and a coal-dominated energy structure. Without reducing the proportion of coal in the energy structure and the total amount of coal consumption, the positive impacts of many innovations and technologies applied at the consumption end would be limited. For example, the innovation of electric cars and related infrastructure would be meaningless if the cars were powered by electricity generated from coal. It might ease China's reliance on imported oil, but there would be little reduction in CO_2 emissions and urban air pollution. Therefore, to study China's energy structure and climate change mitigation, two topics must be discussed, the conventional coal energy and the renewable energies.

Closure of small inefficient thermal plants

Small-scale inefficient and highly polluting coal-fired power plants in China had a strong connection to the country's rapid economic development. China constantly experienced energy shortages since the reform and open up in the late 1970s. In response to this situation, the National Planning Commission (the predecessor of National Development and Reform Commission [NDRC]) announced the *Tentative Regulations on the Development of Small Thermal Power Plants* in 1986. However, China continued to experience energy shortages. In fact, the only periods that China did not suffer from energy shortage were during the years of global financial crisis, when China's economy was impacted by global economic downturn (Lin 2007, p. 916). For instance, 1998 to 2001 were the rare four years that China could basically satisfy its energy demand due to the influence of the 1997 Asian Financial Crisis. However, severe energy shortages bounced back in 2002 when the financial crisis eased (Zhou, Levine & Price 2010, p. 6439). To relieve severe energy shortages, local governments were encouraged to seek funds independently to build small-scale coal-fired power plants with a capacity lower than 25 MW. These small coal-fired power plants were financially independent, and they were therefore administratively independent. Due to these financial and administrative freedoms, local governments put great effort into the development, and small thermal plants flourished across the country (Qi 2013, p. 127).

Small-scale coal-fired power plants might have temporarily relieved energy shortages, but they have caused enormous detrimental impacts on the environment such as severe atmospheric pollution and global climate change.

Generally speaking, small-scale coal-fired power plants could not meet the efficiency level of large plants and rarely installed pollution control equipment; therefore, they were energy and emission intensive. According to the Director General of the National Energy Bureau, Zhao Xiaoping, to generate the same amount of electricity, small plants generally consumed 30–50% more coal than large plants (Weipu 2007, p. 16). In order to ease domestic atmospheric pollution and respond to the international community's call on climate change mitigation, the Chinese government decided to shut down small-scale coal-fired power plants starting from the 11th Five-Year Plan (2006–2010). The closure of inefficient and highly polluting small-scale coal-fired power plants was clearly written in the energy sector ECERS as a major policy (The Central People's Government of PRC 2006a, n.p.).

Regulations and procedures for closures

In 2007, the State Council of PRC approved the *Several Suggestions on the Acceleration of Closing Down Small Thermal Power Plants* jointly issued by the NDRC and the State Energy Office (State Council of PRC 2007, n.p.). It suggested the closure of small-scale conventional coal-fired power plants within the coverage of major grid according to relevant laws, regulations, and the State Council's requirements. The targeted coal-fired power plants included plants with a capacity under 50,000 kW; plants operated over 20 years with a capacity under 100,000 kW; plants which had reached life expectancy with a capacity under 200,000 kW; plants with a coal consumption level 10% higher than the province's average where it was located or 15% higher than the national average; and plants with high emissions which did not meet relevant environmental protection regulations. The government welcomed units with combined heat and electricity generation functions regardless of their size as they played an important role providing heating in northern China during winter. However, their efficiency level should still be monitored. For heat and electricity combined units with a coal consumption level 10% higher than the province's average or 15% higher than national average outside the heating providing period, electricity generation was banned or restricted during the warmer months (State Council of PRC 2007, n.p.).

The implementation of the policy started with NDRC assigning decommission tasks for provincial governments and large energy state-owned enterprises (SOEs). All provincial level governments were required to submit detailed implementation strategy and annual closure plans before 31 March 2007 (China Electricity News Network 2007, n.p.). A target responsibility system (TRS) containing comprehensive information of the targeted small coal-fired power plants was then established. The provincial, city, and district governments played a supervising role in drafting the implementation strategies and the annual closure plans for their localities and made these documents available to the public. The implementation strategy should clearly state the division of responsibility and post-closure details, including worker arrangements and enterprises' debt and asset disposals. The annual closure plan should clearly identify the names of the targeted plants and the deadlines for closure. Energy enterprises were responsible for the direct operation of the closures, which should all be strictly implemented according to the relevant provincial government's implementation plan (State Council of PRC 2007, n.p.).

All provincial governments were required to report periodically to the NDRC and the State Electricity Regulatory Commission (SERC) on the implementation of the closure of small coal-fired power plants within their jurisdictions. For plants which exceeded the closing deadlines, electricity regulation agencies should immediately suspend their licences; grid companies should disconnect them on time and stop purchasing electricity from them; power dispatching centres should stop providing dispatching service; and banks and other financial organisations should stop providing loans. The units should be eliminated straight away after closure and should not be reconstructed in other localities. For small plants which refused to close, the relevant department of the provincial government or above could order immediate closure by suspending their licences and terminating any future development proposal until the targeted small plants were shut down. For enterprises which cheated to avoid closure or those who closed existing plants but built similar small coal-fired plants elsewhere, the governments would order immediate demolition and pursue legal action against the relevant personnel. For local officials who acted beyond their authority for approving the construction of small coal-fired power plants, the provincial government or above would step in to investigate and withdraw the project approval documents (State Council of PRC 2007, n.p.).

Grid companies were also involved in the implementation process. They were required not to provide services to small coal-fired power plants unqualified for continuing operation. The power dispatches of all coal-fired power plants under the coverage of the large grid were required to be supervised by the provincial or higher level power dispatching centres for unified management. The dispatching method should be improved according to the principles of saving energy, protecting the environment, and economic efficiency. Therefore, the power dispatching centres and grid companies were required to give priority to dispatch electricity generated from renewable sources and high-efficiency coal-fired power plants but limit electricity generated from inefficient and polluting units (State Council of PRC 2007, n.p.).

All levels of environmental protection bureaus (EPBs) should also play an important role in the monitoring and auditing of major pollutants of coal-fired power plants. EPBs should supervise the installation of desulphurisation facilities for coal-fired power plants with a single unit capacity above 135,000 kW outside the closure category. Units which had installed desulphurisation facilities but still did not meet relevant environmental standards would not be paid a higher desulphurised electricity price when connected to the grid (State Council of PRC 2007, n.p.).

The implementation of closure of small coal-fired power plants was largely delivered through administrative procedures. It mainly involved the government, large energy SOEs, and large grid companies with minor involvement from the Ministry of Environmental Protection (MEP) and lower levels of EPBs. It had minimum involvement from the public and civil society. However, this command and control style implementation was effective in China's situation and particularly so in the case of closing small coal-fired power plants. The next section will discuss why and how the administrative procedures were effective.

The strengths of closure of small coal-fired power plants

The closure of small-scale inefficient coal-fired power plants in China achieved the set targets during both the 11th and the 12th Five-Year Plans (Sun 2009, n.p.; Wang 2015, n.p.). This was largely due to the strong central control over provincial governments and large energy SOEs and the adoption of the TRS. As mentioned in Chapter 2, China's political system was characterised by a hierarchical unitary structure. This structure, compared with many Western democratic governments, has its major drawbacks. For instance, provincial and local governments hold accountability to higher levels of government rather than the public in their administrative regions (Guo 2007, p. 378). As a result, people's voices are often ignored and are unable to reach the central government. However, in the case of shutting down small coal-fired power plants, this top-down TRS was extremely effective.

First of all, provincial and local leaders' work performance was closely linked to how well they accomplished the tasks that the upper level governments appointed to them. Whether or not they completed the tasks and how promptly and by how much extra they achieved were all important indicators which would be considered for their future political promotions (Qi & Wu 2013, p. 308). Due to this mode of upward accountability, provincial governments divided the provincial task to county governments, and the county governments did the same to the township governments. As a result, the pressure was passed down the hierarchy. In order to pursue one's own promotion opportunity, higher level leaders would pass down pressure to lower level government to better complete the tasks; and majority of lower levels government leaders did their best to please their supervisors in higher levels of governments by shutting down small coal-fired power plants in a prompt manner (Xue & Yang 2011, p. 5).

Second, this top-down command and control system also allowed provincial and local governments to eliminate much public consultation. Big stakeholders such as large energy SOEs received enormous funding and support from the government. In order to maintain a good relationship with the government so as to obtain funding and future development opportunities, they were rarely against the government's policy decisions and the closure tasks the government appointed to them (Williams 2014, p. 12). Workers who were employed by small coal-fired power plants within the closure categories were among the most affected. However, due to the fact that majority of workers' trade unions were located and managed by individual companies, the closure of the small coal-fired power plants also meant the disintegration of the unions. As a result, protests and large-scale strikes led by unions were extremely rare.

According to Interviewee 8, an official from National Energy Administration (NEA), NDRC, the strong command and control style of governance was a key factor for the successful implementation of the closure of small coal-fired power plants.

> We realised our system still had many problems in terms of the implementation of policies, but in the case of closure of small coal-fired power plants, our system had its advantages. It was quick, strict, and effective and results were seen in a short period of time. I believed this could not be achieved in any western democratic countries, given the size of the target and the short time frame.

Last, the closure of highly polluting small coal-fired power plants was generally welcomed by the majority of the public. As atmospheric pollution intensified in China, any measures that could potentially improve air quality were embraced by the local residents. Interviewee 7, an academic from the Chinese Academy of Social Sciences (CASS) put it this way:

> According to my many years of experience doing research at the local level, majority of the general public don't care about who the top leader is and what the major policies are. They want to see the results. As long as the policies bring good outcome to local environment and their livelihood, they'll be generally happy.

Due to the above mentioned reasons, the closure of small coal-fired power plants in China was relatively quick and effective.

Problems emerging during the closure process

China's nationwide closure of small coal-fired power plants during the 11th and the 12th Five-Year Plans achieved the set targets. In this sense, it was a successful policy. However, a range of problems emerged during the implementation process. The unemployment issue was the biggest problem. The dissatisfaction of some private owners of shutdown small plants was also exposed. Last, cheating and fraudulent behaviour of some of the dissatisfied small plants seriously affected the accountability of the policy and diminished the positive outcome of the policy.

Dissatisfaction in terms of compensation and reemployment was a significant phenomenon among workers of closed small plants. The *Several Suggestions on the Acceleration of Closing Down Small Thermal Power Plants* required enterprises to make a worker allocation plan and appropriately settle laid-off workers according to relevant laws and regulations after the closures. Theoretically, an enterprise required to partially close its operations should absorb the affected workers within the enterprise. An enterprise with all its units completely shut down should by law appropriately deal with compensation and social security payments. Enterprises with upgrading tasks or new expansions should give employment priority to workers from closed small plants. Local government should also supervise the allocation of workers (State Council of PRC 2007, n.p.). However, the enterprises with new expansions were mainly large SOEs. The majority of the small non-SOEs did not have the capacity to provide alternative employment opportunities for their workers, and workers' reemployment situation was difficult (Sun et al. 2008, p. 32).

When Xu Yongsheng, the Power Division Director of National Energy Administration, responded live to netizens on 13 May 2010 on the topic of "Accelerating backward productivity and enhancing energy conservation and emission reduction", he stated:

> Being people oriented is an important principle of government work. The worker arrangement of closed small coal-fired power plant was ranked the number one most concerning issue of the government. In order to appropriately arrange 400,000 workers, we adopted several methods. For instance, we allowed the closed plants to sell their electricity generation allowance and pollution allowance so that they can obtain partial financial compensation from this. The central finance also allocated special funds as incentives during the years for closures in some inland and western provinces where the economy faced more difficulties. We also required enterprises and local government to provide training for the affected workers to improve their skills for reemployment. Large energy SOEs were required to lead the internal transfer of workers from closed small plants. Private and foreign owned enterprises were also required to provide worker allocation plans. Closures were only implemented if the worker relocation plan was sound. (The Central People's Government of PRC 2010, n.p.)

There was a major problem with Wu's response to the public inquiry – lack of quantitative information on the settlement of the laid-off workers. He basically repeated the requirements and procedures in the policy announced at the start of the 11th Five-Year Plan (2006–2010) but included little data about the percentage of workers reemployed. There were few official statistics on the reemployment rates or compensation allocations. Cai et al. (2011, p. 5998) conducted a scenario analysis on the impact of closure of small coal-fired power plants on unemployment which indicated that the policy led to a reduction of approximately 559,000 direct job opportunities in the power sector. There were many cases where laid-off workers expressed their dissatisfaction on social networking websites. These indicated that unemployment caused by the closure of small coal-fired power plants were real and more serious than it appeared to be. In the next section, two typical cases from two of China's most widely used public forums, *baidu tieba* and *bbs163*, are reviewed.

Case 1:

Posting on *baidu tieba* (a popular Chinese public online forum) by author "8`1 red collar badge" on 15 January 2016, titled *How many laid-off workers did not receive appropriate arrangements in the Inner Mongolia West Electricity System?*

The author, as one of the 175 laid-off workers from the closures of Hohhot Qingshihe No.1 and No.2 coal-fired power plants, described their prolonged and tough journey in fighting for their rights and justice. The two closed plants announced bankruptcy and dismissed all the workers. The enterprise did not appropriately arrange the laid-off workers according to the government's policy, leaving the 175 workers unemployed and uncompensated. The representatives of the laid-off workers pursued petitions in November 2006, March 2007, June 2007, and September 2009 and visited the following government departments and units to appeal for help: Inner Mongolia Qingshuihe Power Supply Bureau, the People's Government of Qingshuihe Township, Hohhot Power Supply Bureau, Inner Mongolia Electric Power Group Company, the Bureau of Letters, and Calls of Inner Mongolia Autonomous Region Party Committee. Although the majority of these government departments showed sympathy to the workers' situation, they all attempted to shift responsibility to other departments.

The problem was still not resolved by the time the author posted their experience on *baidu tieba*. On 28 December 2015, the 175 laid-off workers once again submitted their petition materials to the Comprehensive Management Department of Hohhot Power Supply Bureau, and waited to see the government's response (Baidutieba 2016).

Cases like this were not rare. In the next case, an unemployed worker from a small factory-owned coal-fired power plant also expressed his/her concerns over compensation and future job opportunities.

Case 2:

Posting on *bbs163* by author "llhhjj760" on 21 August 2010, titled *Frustration, small coal-fired power plant closed, who cares about the laid-off workers?*

The closure of the small coal-fired power plant owned by CITIC Heavy Industries Co., Ltd. (CITIC) Machinery happened on 10 March 2010. The author, "llhhjj760", one of the laid-off workers from the closure

> of this small plant, used the word "frustration" to describe the workers' feelings. llhhjj760 expressed his/her concerns over the unclear compensation and job opportunities after leaving the closed plant. According to the *Labour Laws of the PRC*, closed down power plants were not allowed to force employees to have a seniority buyout (redundancy payment), rather, enterprises should provide partial social security payment to their previous employees. In llhhjj760's case, his/her company just changed the name of seniority buyout to "financial compensation and cancellation of contract", which basically meant a small one-off payment to the employee, avoiding any further responsibility towards them (BBS163 2014). Unemployment caused by the closure of the small coal-fired power plant was difficult for those it affected. Although some commented it was a necessary move towards better energy conservation and environmental protection, the allocation of appropriate social security and training for the unemployed should be well managed.

Dissatisfaction was not only among the laid-off workers who did not get proper compensation, it also existed among private owners of the closed small plants. As mentioned previously, many of China's small coal-fired power plants were established in response to the national call during energy shortages. They were financed by private funds or local government raised funds (Qi 2013, p. 127). As a result, their closures were much more complicated compared with those owned by SOEs because of the diversity of interests and stakeholders. Some coal-fired power plants were established by local governments through the method of social fund raising, and the repayment to the public relied on the continuing production of the plants. The forced closure of such types of small coal-fired plants left many fund creditors bearing financial loss. There were also small coal-fired power plants operated on a shareholding system. The employees of the plants were also shareholders. Once the plants were shut down, the employees were unable to cash their shares (Li 2007, n.p.). The national policy *Several Suggestions on the Acceleration of Closing Down Small Thermal Power Plants* did not clearly state individual methods in dealing with different types of ownership of small coal-fired power plants. As a result, the implementation of closures beyond SOE-owned plants was not smooth and filled with conflicts of interest. A previous owner of a private small coal-fired power plant (Interviewee 9) expressed his dissatisfaction towards the closure of his company:

> A few years ago, when the plant was first approved, it was legal and all the equipment was said to be good. The government at that time encouraged my investment. But how could they suddenly require me to shut it down! I put in so much effort running it. It was my career. Although I received compensation, it was nowhere near what I had devoted to the company. I also felt sorry for my employees as I don't have the capacity to offer them new employment opportunities. They are doing it tougher than me.

Unemployment and unfair compensation to laid-off workers cause serious social issues. Although it may not yet cause severe social instability, it has the potential to do so. Most important of all, it severely affected the livelihood of laid-off workers and made their daily life a struggle. Therefore, how to appropriately solve this problem deserves much attention, and some recommendation will be provided in Chapter 6. Also, since the dissatisfaction was not uncommon among the closure of non-state-owned small coal-fired power plants (Li 2007, n.p.), some small coal-fired power plants sought ways to exploit the policy loopholes to avoid closure; while others even secretly reopened the plants after closure.

One of the biggest loopholes in the policy of closure of inefficient small coal-fired power plants was the privilege for combined heat and electricity generation units. The government treated the combined-function units differently from the single-function units. The closure decision was generally made through efficiency monitoring rather than the units' size. Due to this reason, some single-function small coal-fired power plants, which would have otherwise been on the closure list, converted their units into combined-function units to avoid closure. They generated electricity business as usual and only limited the generation during inspections. Due to the lack of resources of local EPBs, the inspections were not frequent. As a result, the combined heat and electricity generation function had been used as a cover-up by some small coal-fired power plants for the purpose of continuing their operation (Cheng 2009, p. 108).

There were also small coal-fired power plants secretly reopened after the closure. According to Zhou (2010, p. 270), this phenomenon was not rare. Some small coal-fired power plants announced closure but in reality never stopped their operations. This was partially due to the relevant government department's lack of seriousness during the implementation process. As Li (2007, n.p.) pointed out, the implementation of closing down small coal-fired power plants was carried out by local electricity providers under national grid companies in some regions. Their typical implementation method was to simply put a closure sticker on the boiler, and that counted as the completion of the closure. Such implementation lacked the authority and seriousness, and as a result, many small coal-fired power plants did not take it seriously and continued to operate as usual. According to Ferris and Zhang (2015, p. 68), another reason that some small coal-fired power plants dared to take the risk to reopen was that they had special connections with local governments who gave them the green light to do so. Interviewee 10, the head of an influential ENGO, discussed this problem with great concern. Interviewee 10 stated:

> Under such circumstances, even local EPBs had no way of dealing with the violation because local governments were their direct supervisors. The illegal operation after the officially announced closure was disguised due to the far reach of the central government. We recently investigated

> one such case. A small coal-fired power plant "shut down" three years ago was still in operation. The person in charge responded to our enquiry as that the facilities were to be sold to another country and they were just providing a demonstration to their clients as required.

Cases such as the one described by Interviewee 10 needed to go through a legal process in order to be solved. It required social organisations and the local public to be brave enough to get involved. However, due to a range of limitations on China's civil society and people's traditional belief that "ordinary people should not challenge the authority", environmental lawsuits towards the cheating behaviour of some small coal-fired plants were uncommon.

Another issue of relevance to this book is that majority of fraudulent and cheating behaviour was observed in small coal-fired power plants owned by non-SOEs. For small coal-fired power plants operated under major energy SOEs, the demolition of the facilities was usually on site, on time, and strictly according to the high-level government's guidelines. This reflected different characteristics of these two types of ownership enterprises in terms of pursuing profit. Non-state-owned small plants obviously received less financial support from the central government; as a result, some undertook risky actions to maintain their profit, while large energy SOEs disdained to maintain small plants as they had better opportunities for future development of large projects. Although China's SOEs were regarded as an inefficient form of economic entity, their less profit-oriented nature actually helped in the small coal-fired power plants closures and contributed greatly to the successful implementation of the policy. Non-state-owned small plants, to the contrary, reduced the accountability of the policy and the officially announced policy outcome figures. Therefore, how to treat SOEs and non-SOEs differently in climate-related energy policies requires careful consideration. This was what the *Several Suggestions on the Acceleration of Closing Down Small Thermal Power Plants* clearly lacked. In this book, however, this issue will be dealt with in Chapter 6 with a potential solution proposed.

Establishment of large-scale efficient coal-fired power plants

The closure of small coal-fired power plants was the precondition of establishing large-scale new plants since the 11th Five-Year Plan. According to the *Several Suggestions on the Acceleration of Closing Down Small Thermal Power Plants*, the closure of small coal-fired power plants was an important assessment criteria for the approvals of new energy projects. The government would give priority to a company's new energy project proposal if the company had closed down a large number of small units and made appropriate arrangements for its laid-off workers. For a new proposal of a 300,000 kW coal-fired power plant, the company's combined small plants closure capacity must have reached 80% of the new capacity, which was 240,000 kW, so that the proposal could be included in the national electricity development plan. Similarly, the condition of a new 600,000 kW unit was that its combined small plants closure capacity exceeded 70% of the new unit, whereas a new 1,000,000 kW unit required a closure capacity of small plants which reached 60% of the new unit (State Council of PRC 2007, n.p.).

Closure of inefficient small plants as a precondition of establishing large-scale coal-fired power plants was often ignored by the Western media who simply emphasised the fast speed and big size of China's coal-fired power plant development (Stanway 2014, n.p.), which was misleading. The coal energy development under ECERS was not a blind expansion. Although the total capacity of coal energy slightly increased, the overall emissions were offset by the increased efficiency of the new units. In this sense, ECERS was effective in reducing energy intensity and carbon intensity in China's coal energy sector. However, large coal-fired power plants, although more efficient compared with the closed small ones, still could not avoid the environmental impacts associated with coal burning. In fact, the larger scale coal-fired power plants appeared to have more intense impact on local environments. Numerous complaints of local residents on newly established large-scale coal-fired power plants appeared on the Internet across diverse geographical locations.

In Guangdong province, the number of residents' complaints about Zhongshan Coal-fired Power Plant increased since December 2013 due to the expansion of 2 × 300,000 kW heat and electricity dual-function units. The expansion was located in between Huangpu and Nantou townships. It was under the energy sector ECERS which aimed at replacing small coal-fired power plants with large ones. Zhongshan city government received eight complaints within the first six months for the worsening air pollution.

On 22 July 2014, a nearby resident with the family name Chen complained about the air pollution:

> We are residents of Jinxiu Dongfang residential district which is just under two kilometres from Zhongshan Coal-fired Power Plant. We often smell an unpleasant smell. The smell is especially strong at nighttime. We request the government to conduct an investigation into the new expansion of Zhongshan Coal-fired Power Plant.

Another resident from the same residential district reflected that the air pollution was getting worse. The residents dared not open their windows at night. The smell of coal-burning was very strong and black smoke was often seen from the chimney of the power plant. Some residents also mentioned that tough rain stains remain on their cars when parked outside the apartment. The stains were so difficult to clean off that they suspected it was the result of the acid rain.

In response to the resident complaints, the company published a letter on the official website of Huangpu township. According to the letter, the

Zhongshan expansion project started in July 2011. Its environmental impact assessment gained approval from MEP and final approval from the NDRC. The project was completely legitimate and was encouraged by the central government. The company had installed highly efficient dust and sulphur control facilities in 2012 and had already passed the completion assessment of EPBs. The company was within the state-monitored category which meant it had an online monitoring system installed within its unit chimney and received 24-hour monitoring from the EPB; therefore, emissions from the chimney were guaranteed to meet the national standards. The company adopted a completely sealed facility which eliminated 99.5% dust, and the desulphurisation rate and denitration rate reached 99% and 89%, respectively. This procedure guaranteed the sound environment around the power plant. The expansion of Zhongshan Coal-fired Power Plant had replaced 127 small units in four townships. The sulphur dioxide emission in the region had dropped from 9700 ton/year to 700 ton/year. Therefore, according to the company's statement, the expansion project had largely reduced air pollution and significantly improved air quality in Huangpu and Nantou townships (Cao 2014, n.p.).

The contradictory statements between the local residents and coal-fired power companies with coal-fired power plant expansion projects were not rare. In Heyuan city, hundreds of residents gathered together to oppose the development of Stage-II of Heyuan Coal-fired Power Plant. The residents complained about the deteriorated air quality and discomfort in the throat and lung since the completion of Stage-I Heyuan Coal-fired Power Plant. Liu Dongbin, member of the National Committee of the Chinese People's Political Consultative Conference (CPPCC), also stated that the air quality excellent days dropped dramatically from 365 days in 2011 to 129 days in 2014 during the Heyuan city two sessions[1]. Heyuan Coal-fired Power Plant emitted nearly 30,000 tons of sulphur dioxide and nitrogen oxide since production, which was a significant contributor to the more frequent and worsening smog problem. However, according to an official document of Heyuan city EPB, from 2011 to 2014, sulphur dioxide, nitrogen oxide, and soot emissions of Heyuan Coal-fired Power Plant all met national standards and showed a decreasing trend year by year (Shenzhen City News 2015, n.p.).

The contradictory statements between local residents and official responses over the pollution of coal-fired power plant expansions raised suspicions as to whether bigger and more efficient coal-fired power plant really generated less emission or it just seemingly did so. According to the public relations representative of Zhongshan Coal-fired Power Plant, all the company's pollution control devices adopted advanced technology and guaranteed minimum emissions; however, whether or not the devices were used continuously was in doubt. As a public servant of EPB (web ID: Lantian baiyun zhongkeqi) pointed out on Tianya BBS in 2016, the fundamental reason for China's serious air pollution was not the lack of advanced pollution control devices but companies purposely avoiding using them to pursue maximum profit, and this phenomenon was extremely common across the country. Even though

EPBs could computer monitor pollution levels through devices installed into some power plants' chimneys, the power plants could avoid using chimneys with the monitoring system, and only using the ones without the monitoring system (Lantian baiyun zhongkeqi 2016, n.p.). Although Zhongshan Coal-fired Power Plant claimed that the plant received 24-hour monitoring from the EPB, it was still possible for it to avoid running its pollution treatment devices. This could be done by bribing the local EPB or the local government. Heyuan city EPB's official response to the pollution problem of Heyuan Coal-fired Power Plant also attracted suspicion, as the local residents' complaints and the CPPCC member's data all indicated worsening air pollution after the completion and running of Heyuan Coal-fired Power Plant. Nonetheless, if the coal-fired power companies did strictly implement all pollution control procedures according its statement and the EPB did provide an accurate report, then the conclusion would be quite different from the expectations. It would indicate that big and more efficient coal power plants still produced large amount of environmental pollution and brought much negative impact to local residents.

The pollution caused by large coal-fired power plants was not limited to air pollution. The improper storage of coal also causes contamination to the surrounding environment. In Changchun city, Jilin province, the manager of Yuquan Bean Product Co., Ltd., Fang Wenjie, complained that each day almost half a ton of dried bean curd had been dyed black due to the coal dust drifting from the nearby Guodian Jilin Hualong No. 1 Coal-fired Power Plant since 2014. The problem had not been resolved for a prolonged period. Fang approached the city and provincial EPBs multiple times but received no clear response. He got extremely frustrated and decided to seek help from the public media. After the media got involved, the provincial EPB required Guodian Jilin Hualong No. 1 Coal-fired Power Plant to manage its coal stock within a sealed complex within a time limit; however, the "time limit" was not clearly defined (Hua 2015, n.p.). There is no doubt that coal-fired power plants bring multiple types of environmental pollution and impact on people's livelihood; therefore, the key to minimise such impact is to reduce the production and consumption of coal-generated electricity.

The current reality is, unfortunately, China's national energy consumption is still rising. And the proportion of China's coal electricity did not decrease significantly in the national energy structure. By the end of the 12th Five-Year Plan (2011–2015), coal energy still took up 67% of the total national energy supply. Although only increased slowly, by 2015, China's coal consumption reached a historical peak of 4.3 billion tons (Wang 2016, n.p.). This is to say, to replace small coal-fired power plants with larger ones alone was not enough to combat the associated pollution and climate change consequence. ECERS in the energy sector fulfilled its energy conservation function, but the emission reduction function was still limited. In order to fulfil the emission reduction task of ECERS, there has to be a bigger proportional increase in renewable energies – big enough in scale to replace coal energy.

Renewable energy under ECERS

New and renewable energy development under ECERS achieved positive outcomes in terms of growth speed and scale during both the 11th Five-Year Plan (2006–2010) and the 12th Five-Year Plan (2011–2015). During the 11th Five-Year Plan, China's electricity generated from new and renewable energies (including nuclear energy) reached 880 billion kWh in 2010 which was a 91% increase compared with the 460 billion kWh in 2005. It increased the proportion of new and renewable energy generation to 21.1% of the nation's total energy production (Liu GY 2011, p. 11). According to the statistics of the National Energy Bureau, China's new and renewable energy also experienced rapid development during the 12th Five-Year Plan. The capacity of nuclear energy, hydro energy, wind energy, and solar energy increased 2.4 times, 1.6 times, 4 times, and 168 times compared with the 11th Five-Year Plan achievements (CCTV News 2015, n.p.). At the November 2015 Paris Climate Change Conference, President Xi Jinping displayed China's achievements in new and renewable energies. By the time of the conference, China's installed capacity of renewable energy took up 24% of the world's total, and newly installed renewable energy capacity took up 42% of the world's new growth, which made China the world's number one nation in new and renewable energy development (China Daily 2015, n.p.). These numbers were significant; however, the quality and efficiency were often disguised by such glorious numbers. The most significant problem in China's renewable energy development was the wastage of renewable energy.

In May 2010, China's State Council announced *Several Suggestions in Encouraging the Healthy Development of Private Investment* (State Council of PRC 2010, n.p.). In early 2012, NEA and the SERC published the detailed implementation methods in regards to the *Suggestions*. Following this, *China Economic Times – Energy Weekly* spent two months conducting numerous interviews with private enterprise CEOs, SOE senior managers, and experts and consultants in the renewable energy field for a purpose of realistically displaying the status of China's private enterprises in the renewable energy sector and seeking suitable channels for private investment to enter the industry. From the interviews it was found that many private renewable energy investors in fact faced incredible challenges and struggled to survive (Zhang 2012, n.p.). These results echoed the interviews conducted for this book: China's renewable energy industry developed fast at a relatively significant scale; however, the quality and efficiency of the entire industry was in question, and the private investors faced great challenges. Interviewee 14 from CASS explained this situation as follows:

> China's renewable energy development, especially wind and solar power, is in a blind rush. So many projects were approved and large amount of investment was put in. But in the end, many of them were not connected

> to grids. The ones that did connect to grids were often not allowed to operate at full capacity. Many of the investors experienced enormous financial difficulties.

The slowdown of the global and Chinese economy determined the decreased market demand for wind turbines and solar panels. In China, private investment performed a leading role in many renewable energy related manufacturing industries, and many of them experienced profit deficit (Shi 2012, p. 21). The major carbon trading body – the European Commission clearly indicated the prohibition on using new-project CERs unless they are from the least developed countries in Article 11a (4–5) (European Commission 2012, n.p.) Obviously, China was not in this category. Without further stimulation of CDM, domestic private investment in renewable energies could not maintain its previous volume. The chain reaction passed onto the renewable energy related manufacturing industry forcing enterprises to either reduce productivity or be pushed out of the industry due to serious competition (Shi 2012, p. 21).

The dismal situation of the related manufacturing industry was fundamentally determined by the less healthy status of China's renewable energy industry. According to the 2012 semi-annual report of renewable energy listed companies, the whole industry faced survival challenges. Nearly 80% of wind energy enterprises experienced profit deficit. The solar energy even faced entire sectoral deficit (Zhang 2012, n.p.).

Unfair competition between large energy SOEs and small private investors of renewable energy

Private investors in renewable energies were doing it even tougher as they faced more challenges and obstacles compared with renewable energy companies under large energy SOEs. According to NEA data, private investment in new and renewable energies exceeded 800 billion yuan (approximately US$130 billion) in 2012, and its energy production capacity exceeded 49 million kW which took up 18% of the national new and renewable energy capacity (NEA 2012, n.p.). This percentage was not a small number when considering China's entire renewable energy investment; however, these investors were not guaranteed a fair competitive environment.

First of all, the government was unwilling to provide a specific policy to lower risks for private investors even though the capital requirement in the renewable energy sector was high. Zhao Changwen, senior researcher and Director of the State Council Enterprise Development Research Centre confirmed this point when he was interviewed by *China Economic Times – Energy Weekly*.

> The government should provide a fair competition environment for private investors in the energy sector. However, it's up to the investors to conduct long-term strategies and risk assessment. It is a market behaviour

> for private capital to enter the energy sector, therefore, investors should bear their own risks and losses. It is not recommendable to interfere with private capital gains through policy intervention.

In regards to the government's claim of the fair competitive environment, Interviewee 11, a private owner of a small-scale wind farm, expressed his strong disagreement and concerns. He believed the seemingly fair policies, such as price sharing of higher renewable energy cost, had in fact disguised the unfairness between private small investors and large energy SOEs:

> Some large energy SOEs accept renewable energy projects purely for the purpose of boosting their bargaining power with the government for future traditional energy project opportunities which are of course cheaper and more profitable. They don't care about profit margins or even profit deficit from these renewable energy projects because they have profit from large coal and gas projects to compensate this. Also, investing in renewable energy makes them look good. Meanwhile, we [small private investors in renewable energies] don't have government financial support; we don't have coal or gas power to boost our overall returns; we work hard but we are totally on our own. There is no fairness and we have to cope with it.

What Interviewee 11 described was true as renewable energy projects were indeed not large energy SOEs' focus; their focus was instead traditional thermal and nuclear energy (Liu & Kokko 2010, p. 5525). According to Wang (2010, p. 706), all large energy SOEs were required by the government to include a certain percentage of non-hydro renewable energy in their portfolio. Moreover, the government's concession wind energy projects were mainly obtained by large energy SOEs. By the end of 2008, SOEs had taken 97% of the central government's total concession wind energy projects. Only one private company, Farsighted Investment Group Company, had ever won a small concession wind power contract (Liu & Kokko 2010, p. 5525). This was to say, non-SOEs had been in a weak position at the investment stage when competing for the government's favourable concessions. This situation reflected the hidden unfairness between private investors and energy SOEs in terms of investment conditions and the competition environment.

Private renewable energy investors not only faced competition from large energy SOEs but were also threatened by the new competition from large grid companies. According to Interviewee 12, a consultant from Saidi Zhiku Consultancy, one big problem that renewable energy faced was the difficulty in connecting to the grid. Although the government required grid companies to commit full purchase of renewable energies and grid companies did invest large amounts of money to establish a new grid to accommodate renewable energies, there were still 20–30% of wind power projects not connected to the grid (Qi 2013, p. 157). In 2015, the total wasted wind energy proportion in the northwestern province of Gansu reached 43%. The wind power failing to connect to the grid nationwide was sufficient to power Beijing for the whole year of 2015 (South China Morning Post 2016).

The monopoly nature of China's grid was the biggest obstacle for private renewable energy investors to enter the market and play their role. The investment of two major state-owned grid companies in wind energy had made things worse for small private investors. Interviewee 13 explained this situation as follows:

> In June last year [2012], Southern Power Grid Holdings announced that they established Southern Offshore Wind Power Joint Development Co., Ltd. State Grid had done this before Southern Power Grid. This can potentially cause unfair competition. We all know that competition is serious in the wind energy sector, and the grid can only accommodate a certain amount of energy generated by wind. Which wind farms would big grid companies connect first? Their own or some irrelevant small wind farms? The answer is obvious.

The vulnerability that small private investors endured when competing with large energy and grid SOEs was not a good sign for the general efficiency and health of the renewable energy sector. As Chen et al. (2011, p. 263) pointed out, the sensitivity of investment expenditure of Chinese SOEs was weaker than that of non-SOEs. Investment efficiency decreased dramatically if the SOE had special political connections with the government. This echoed the view of Interviewee 11 that large energy SOEs were not really concerned about the efficiency and profitability of renewable energy projects. Ironically, the less efficient and less caring SOEs got most resources from the government, while the private investors who genuinely cared about efficiency and profitability received little attention. This seriously contradicted the principle of the market economy. Chapter 6 will propose some recommendations to address this issue. Nonetheless, the unfair competition small private investors faced was not yet the biggest weakness of China's renewable energy industry. The real problem was the waste of resources and the associated administrative and policy barriers, especially in the wind energy sector.

Waste of resources in the wind energy sector

In 2015, the three regions including northwest China, northeast China, and north China abandoned a combined 33.9 billion kW of wind energy (NEA 2016, n.p.). According to the NEA data, the national average utilisation of wind in 2015 decreased to 1728 hours which was 165 hours less than that in 2014. The abandoning of wind power reached a historical record (NEA 2016, n.p.). This large-scale abandoning of wind power happened when wind resources were plentiful and wind turbine systems were in good working condition. In response to this problem, NEA officially stated that the limited local energy consumption capacity was the major cause of abandoning of wind power in these regions due to less developed economies and a smaller population (NEA 2016, n.p.). Interviewee 1 also mentioned that the unstable characteristics of wind power contributed to the problem: wind farms produced peak load of power over winter months and were stronger during night time, so that the current grid system was incapable of handling it.

The above explanations echoed the statements of many grid companies. However, these factors should not constrain the generation of wind power to such extent in China's northern provinces. According to Qin Haiyan, the Secretary General of Wind Power Committee of Chinese Renewable Energy Society (CRES), wind energy did not require as much peak modulation control as some argued; therefore, the so-called "unstable characteristic" should not be the fundamental reason for the abandoning of wind power. This is because peak modulation service is only required when unexpected incident happens causing sudden energy production and consumption imbalance. For instance, a sudden heat wave which pushed up electricity consumption for cooling, a working electricity generator suddenly broke down, or a temporary stoppage of an energy intensive factory, these were situations that required peak modulation control and intervention such as increasing or reducing generation from certain generators. Although elevated, 80-90% wind power production is predictable. The only part that requires peak modulation is the prediction deviation which would be quite minor. This is to say, as Qin explained, wind energy production could be easily incorporated into the entire energy generation plan as it was not an unpredictable random energy. It was totally misleading to compare wind energy production to some sudden and unexpected incidents which could break the daily planned electricity balance. Furthermore, wind turbines did not work simultaneously on a wind farm. The larger the wind farm, the more this was so. A 1,000,000 kW wind farm would generate an average of 300,000–400,000 kW electricity during 95% of time of the day, which means a 500,000 kW grid allowance would guarantee accommodation of the electricity generated by a 1,000,000 kW wind farm. Therefore, the generation of wind power took up much less grid resources than people thought (Qin 2016, n.p.).

In China, the true reasons underpinning the so-called "limited local consumption capacity" and "incapability of the grid system" were indeed competing interests among renewable energy companies, coal-fired power plants, grid companies, provincial governments, and the central government. The first reason that wind power could not be fully consumed in China's northern provinces was due to the competition from newly added coal power in the same region which created electricity supply surplus. According to the *Electricity Industry "the 12th Five-Year Plan" Research and Planning Report* conducted by China Electricity Council (CEC), the strategic planning of China's coal power was to be concentrated in the "Three Norths" region (Northeast, Northwest, and North China) from 2011 to 2020 (CEC 2011, n.p.). According to CEC, it was a strategic plan due to the fact that the majority of China's coal reserves were deposited in the northern regions. Establishing new coal power plants close to coal mines would significantly reduce transport costs and relieve pressure on the rail system. During the 12th Five-Year Plan (2011–2015), 66% of China's new coal power plants would be located near China's large coal mine bases and 63% during the 13th Five-Year Plan (2016–2020). Since 14 out of 16 of the China's large coal mine bases were in China's "Three Norths" regions, the "Three Norths" regions would be the main area for hosting China's new coal-fired power plants (CEC 2011, n.p.).

Although wind power and coal power expanded in the same region during the same period, coal-generated electricity had always been smoothly connected to the grid, purchased, and consumed, and there was not a single newly established coal-fired power plant forced to disconnect due to "limited local consumption capacity". This obviously contrasted to the situation of wind power in the same region. The competition between coal power and wind power in China's "Three Norths" regions underpinned the fundamental problem in China's electricity planning. Fast and large-scale development of both types of energies within the same region was bound to create a winner and a loser as the energy consumption capacity was limited due the less advanced economy. The facts showed that coal power was the winner and it put a great restraint on the sustainable development of wind power in this region.

This win-lose scenario between traditional coal power and renewable energy reflected Interviewee 11's statement that large energy SOEs attached more importance to traditional power projects and their words weighed much heavier than those of private investors. In the large-scale abandoning of wind power in China's "Three Norths" regions, private listed companies suffered the most. Yinxing Energy Co. Ltd., for example, was a joint-stock company in Ningxia Autonomous Region, whose core business was renewable energy. Within the last quarter of 2015, Yinxing Energy suffered a great net profit deficit of 100 million yuan (approximately US$17 million). The net profit deficit was directly caused by the Autonomous Region Government's administrative restrictions on wind power generation due to the so-called limited grid carrying capacity. To deal with the profit deficit, the major shareholders admitted that they had broken their word in achieving the net profit target due to unforeseen and uncontrollable policy and regulation changes, and had to start drafting a cash compensation plan to shareholders (Zhou 2016, n.p.).

The restrictions on wind power appeared fatal to the survival of many wind power enterprises. Forced to shut down a certain percentage of wind turbines, they would suffer a negative impact on the revenue performance. According to Interviewee 13, an analyst specialising in renewable energies,

> The expenditure of wind farms mainly consisted of depreciation cost and finance expenses, which were basically fixed. There wouldn't be significant elevation from quarter to quarter, as a result, if the provincial government purposely restricts wind power generation, the losses of wind farms would be significant.

In 2015, wind power enterprises in Ningxia were forced to abandon an average 13% of their wind energy production, which wasted a total of 1.3 billion kWh of electricity and wasted 37 times more than in 2014 (NBSC 2016, n.p.). Yinxing Energy also identified that the government restriction on wind power production contributed greatly to its large deficit in the last quarter of 2015. Since October 2015, the government of Ningxia Autonomous Region had used coal power as an engine to drive up the coal mining sector, and leading coal power enterprises had signed a 30 million ton coal purchase contract with the local coal mines. Meanwhile, the Ningxia Economic and Information Technology Committee announced an administrative order to wind energy enterprises to restrict their energy production by up to 60%; in December, the restrictions on some wind energy enterprises' production reached up to 90% (Zhou 2016, n.p.).

The provincial government's preference for coal power was obviously a main contributor to the dismal situation of Ningxia's wind energy industry. To clear this obstacle which restricted the healthy development of wind power, it would require the central government to review the effectiveness of current renewable energy policies and the implementation of the renewable energy law to contain provincial governments' independent decision-making in terms of restricting renewable energy production. But so far, the only action the NEA carried out in dealing with the wastage of wind energy was to ban new wind farm development in Inner Mongolia, Jilin, Heilongjiang, Gansu, Ningxia, and Xinjiang in 2016 (NEA 2016, n.p.). There is no detailed policy in place for the existing wastage problem of renewable energies.

According to Interviewee 14, the lack of effective policy showed the complex nature of the problem. "It's not that NDRC doesn't want to solve the problem, indeed they do not have any good ideas or effective ways to do this. It is simply too complicated a problem involving interests of too many parties". This echoed research conducted by the European Wind Energy Association (EWEA) on the topic of integration of large-scale wind power into the European energy supply system: whether or not a significant amount of wind power was accepted by the European power system depended more on the economics and regulations than technical constrains (EWEA 2005, p. 2). This argument also applies to China where the conflict of interests and regulatory barriers are significant.

The second obstacle that restricted the sustainable development of China's wind power was the disagreements and compromises between the central government and the provincial governments on the planning of wind power. Gansu Jiuquan Wind Power Base was one such example in which the provincial government challenged the central government's planning decisions for the purpose of more economic gain. Jiuquan city of Gansu province was well known for its abundant wind resources, which attracted a range of wind turbine manufacturers for its 5,560,000 kW wind power project. After the project was completed, the manufacturing enterprises were ready to move out due to limited market in Jiuquan. However, the Jiuquan city government and the Gansu provincial government both attempted to stop their move by promising more wind power projects. This was because manufacturing industries had always been big tax payers and contributed a significant proportion to provincial and local governments' tax income. However, this decision was contradictory to the central government's planning. The central government had no intention of approving further wind projects in the Jiuquan area. No one knows how the Gansu provincial government lobbied the central government for more wind power projects and the details of the negotiation. Nonetheless, the Gansu provincial government was partially successful in these negotiations, in that the central government allowed them to build a further 3,000,000 kW wind farm but imposed the condition that the electricity generated from the new project must be consumed within the Northwest grid. This condition later became the focal argument of who should be responsible for the wastage of wind power (Qin 2016, n.p.).

The extra wind power capacity built beyond the central government's original planning, although increased the load for the Northwest grid, should not have led to dramatic restriction and wastage of wind-generated electricity, if all relevant actors had followed the *Renewable Energy Law* and gave priority to accommodating renewable energies. Therefore, the third obstacle for wind power in China was the violation of the *Renewable Energy Law*. In 2015, energy consumption in Gansu province decreased due to the overall slowdown of its economy. However, against this background, there were still a large number of non-renewable energy projects completed and commenced producing electricity. Many existing coal-fired power plants converted into dual-function plants for heating and electricity generation for survival. Decreased market demand and severe competition from other energy sources put extra pressure on wind power. There was a lot of lobbying of the government from vested interests such as the coal mining industry, grid companies, traditional coal-fired power plants and other non-renewable energy production actors (Fenghuang Finance 2015, n.p.). The wind power industry was obviously a loser in this competition. According to Qin Haiyang, cases of local governments violating the *Renewable Energy Law* were numerous. For instance, Yunnan Provincial Industry and Information Technology Commission issued a directive in November 2016 requiring wind energy companies to share a proportion of their profit with coal power companies if they were to continue production. The Ningxia and Xinjiang autonomous region governments issued direct administrative orders to shut down wind farms. These behaviours obviously violated the *Renewable Energy Law* and the principle of energy structure optimisation. For the Jiuquan case, if the provincial government and grid companies obeyed the *Renewable Energy Law*, the wind energy utilisation would have reached 1900 hours; however, in reality, the Jiuquan wind power production was limited to 1200 hours in 2015 which was an administrative restriction of 40% of its production capacity (iChina Energy 2016, n.p.). Therefore, how to guarantee the implementation of the *Renewable Energy Law* and introduce clearer and tougher penalties for violating it should be two issues of urgency for the central government. These issues will be addressed in Chapter 6.

The last factor that contributed to the wastage of wind power was the difficulty of selling wind power to other provinces. China's big energy producing provinces were concentrated in the "Three Norths" regions; however, the economies of these provinces were relatively less advanced. As the nation's overall economy slowed down, many of these provinces had an energy surplus; therefore, it was very difficult to sell electricity to surrounding neighbours (Interviewee 14, 2016). Energy demand of the eastern and southeastern provinces was higher compared to the northern provinces due to their more advanced economic status. But the overall slowdown of the global and national economy meant that these provinces were basically self-sufficient in terms of energy supply (Interviewee 14). Even if local energy could not meet demand, the provincial and local governments would prefer to build new power plants within its administrative area in order to boost local economy and employment (Wu 2008, p. 327). Furthermore, according to the economic analysis conducted by Liu et al. (2014, p. 63), apart from Xinjiang Uygur Autonomous Region, trans-provincial renewable energy trading in Gansu and Ningxia was not economical with only a limited feasible market. So the questions that should be asked now are as follows: Is exporting a feasible scenario for solving the wastage of wind energy in China's northern regions? Are there other possible solutions? These questions will also be addressed in Chapter 6.

All in all, the real problem of China's renewable energy industry was the less optimistic efficiency disguised by the overall development speed and scale. There were numerous factors which contributed to this problem. The central government purposely created a "favourite child", the large energy SOEs, which made it difficult to establish a fair market environment. The traditional coal power and coal mining industry actively lobbied provincial and local governments to maintain their existing interests, which pushed aside the relatively new non-hydro renewable energies in the marketplace. Some provincial and local governments purposely violated the *Renewable Energy Law* and used administrative orders to restrain the operation of renewable energy companies, which sent many renewable energy companies into bankruptcy. President Xi Jinping proposed that by 2030, China's non-fossil fuel energy and fossil fuel energy ratio should be 1:5. China is working towards this goal by building up its renewable energy capacity. However, a large proportion of the capacity was not utilised and a large amount of renewable energy was wasted. There is no point building up the capacity but leaving it unused as it is a waste of money and resources. To fundamentally solve the problem of the wastage of renewable energy, China needs to review its current policy approach and work out two major strategies. First, how and where will renewable energies be consumed? Second, what are the appropriate penalties and how can the penalties be effectively implemented for the violation of the *Renewable Energy Law*? These will be some of the focal points of Chapter 6. In the following section, analysis will be continued focusing on ECERS for energy consumers.

ECERS for energy end users

China is still in the process of industrialisation and urbanisation, and as a result, has a high reliance on energy-intensive products for its major infrastructure. Steel, cement, and building materials produced by heavy industry all consume a large amount of energy. Some of China's light industry, such as textiles, as a major exporting good, also consumes a significant amount of energy (Liu, Kong & Gao 2008, p. 1290). Therefore, ECERS also aimed to improve energy efficiency in these big energy consumers. The majority of China's energy conservation and emission reduction programs covering these big energy consumers were implemented through administrative measures. It was very similar to the previously mentioned coal power sector. It started with the central government setting up national energy intensity reduction targets for each sector and allocating reduction tasks for all provincial districts (State Council of PRC 2006, n.p.). Following this, each provincial government distributed responsibilities to high-energy-consuming enterprises within its administrative territory, especially within large SOEs which had the potential to reduce energy intensity through technological innovation. For medium- and small-scale energy-intensive enterprises which could not pass cost and benefit analyses for technological upgrades, the government would issue them administrative orders for closure and provide relevant compensation (Qi 2013, pp. 224–230).

The overall effectiveness of the administrative approach

Again, similar to the coal power sector, the implementation of these administrative measures was relatively smooth. This was partially due to the effectiveness of China's TRS for officials, reinforced by attractive financial rewards. The implementation results of ECERS were important indicators for the annual performance review of personnel in charge of the large SOEs and tied to local and provincial government officials' promotion decisions (Qi 2013, p. 280). As a result, senior provincial and local officials as well as responsible persons of large SOEs who wished to be promoted in their political career, and most of whom did, would make sure their energy conservation and emission reduction targets were met. Moreover, the government allocated sufficient financial fund for technological upgrades of the enterprises, which was a great incentive for them to actively conduct their energy saving projects. In August 2007, the *Temporary Management Method for Energy Saving Technological Upgrade Financial Reward Fund*, issued by NDRC, stated the energy saving cash reward standards for both the eastern and western provinces. The enterprises in the eastern provinces were to be rewarded 200 yuan (approximately US$33) for saving each ton of coal equivalent from installing or upgrading energy saving equipment. For western provinces, the amount was 250 yuan (approximately US$42). In June 2011, NDRC raised the cash reward standards in the *Management Method for Energy Saving Technological*

Upgrade Financial Reward Fund to 240 yuan (approximately US$40)/tce for eastern provinces and 300 yuan (approximately US$50)/tce for western provinces (MF 2011, n.p.). The cash reward was very attractive to the enterprises, especially those with big energy saving potential. According to Shen Hailong, President of China Energy Conservation Enterprise Association, Senior Consultant of Energy Research Centre of NDRC, the energy saving amount of some enterprises was significant which could reach up to tens of thousands of tons of coal equivalent. For instance, an average saving of 100,000 tce, would entitle an enterprise a 30,000,000 yuan (approximately US$5,000,000) cash reward from the government.

Problems with the ECERS cash rewards

The significant amount of cash reward did encourage many intensive energy consumers to take action towards energy conservation and emission reduction, especially those large SOEs under close supervision of central and provincial governments. However, there were some enterprises conducting fraudulent claims for cash rewards for emission reductions they had not achieved (Hu 2011, n.p.). This cheating behaviour was often observed at the township and village levels far from the reach of the provincial and central governments, where effective auditing was lacking and corruption was common.

For instance, an officer of a county Economic Development Office under the Chongming township, Shanghai, fraudulently claimed over 10,000,000 yuan (approximately US$1,800,000) for a small chemical factory and a small paper mill within the county. Both factories were closed long before due to poor management and outdated facilities; however, the Economic Development Office staff and the owners of the two closed factories attached fake electricity bills to their application and successfully obtained the funds from the government (Cong & Guo 2012, n.p.).

Shandong Province Zaozhuang Cement Waste Heat Utilisation Project fraudulently claimed 8,000,000 yuan (approximately US$1,300,000) for a technology upgrade and 2,450,000 yuan (approximately US$410,000) cash reward without even conducting the project. According to relevant documents, the project located in Zhangshanzi County Houmeng Village had gained the central and provincial NDRC's approvals in April 2008. But it was not until four years later, when the project's business licence was suspended, a small shareholder realised that the project had never commenced (He 2012, n.p.).

These were not the only cases. On 13 May 2011, the Audit Commission announced the 2007–2009 auditing results; 40 enterprises fraudulently claimed energy conservation and emission reduction funds involving a total of 205,000,000 yuan (approximately US$34,000,000) (Chen 2011, n.p.). Although this amount was small compared with the total national input of 124,187,000,000 yuan (approximately US$20,837,400,000), it only represented the fraudulent claims which were discovered by the auditing organisations.

More fraudulent cases that were not uncovered could not be eliminated. The government realised the seriousness of the problem and actively explored more sophisticated auditing options for ECERS during 12th Five-Year Plan (2011–2015).

The introduction of third-party auditing body

During the 11th Five-Year Plan (2006–2010), the majority of enterprises' energy savings were monitored and audited by their own energy conservation management offices. Although enterprises had the choice to conduct energy saving auditing through a third party, around 70% of major enterprises chose to do it by themselves and submitted reports to provincial government periodically. After this, provincial energy conservation authorities would arrange professionals to carry out collective auditing to the received energy conservation reports to evaluate their legitimacy (Qi 2013, p. 280). Once this was completed, cross auditing among provinces would be carried out. For instance, the energy saving auditing department of the Beijing government might audit the enterprises in Guizhou province; and the Guizhou auditing department might audit the energy saving results of enterprises in Shanghai (Hu 2011, n.p.). One problem with cross auditing was the lack of unified standards. Two different auditing departments conducting auditing on the same enterprise might end up with different results. Therefore, it was urgent to introduce a range of qualified third-party auditing bodies with unified national energy saving auditing standards (Hu 2011, n.p.).

On 27 September 2011, NDRC announced a list of 26 auditing organisations. Among this first group of national qualified certifying companies, eight were government institutions, four were accounting firms, and the rest were industry intermediary companies (Qi 2013, p. 281). These third-party auditing bodies performed the second last check up for enterprises' energy saving data. After this was conducted, NDRC, MF would organise an expert committee to perform final review of the third-party auditing results. Despite the fact that the government put efforts to improve auditing capacity and procedures, the number of China's third-party auditing bodies was not yet in proportion to the large number of enterprises with energy saving tasks. Nonetheless, it was a positive start for a more transparent auditing system and would form a more realistic foundation for Chinese enterprises energy saving evaluation. Some suggestions will be proposed in Chapter 6 for improving auditing quality of energy saving and emission reduction. Suggestions on preventing fraudulent claims will also be given in Chapter 6.

EPC in China

Apart from introducing the third-party auditing organisations, the Chinese government also actively experimented with market options to encourage enterprises' energy saving and emission reduction. EPC was the market mechanism that the Chinese government promoted. EPC originated in Western developed countries in the 1970s which utilised the market to promote energy saving through a technological upgrade and advanced energy management system. It usually involves an Energy Service Company (ESCO), which tailors an energy saving strategy for its clients (usually inefficient energy users). After the project is implemented, the client will repay ESCO's investment and cover its profit, with the funds obtained from energy saving. Therefore, EPC in this sense is an investment method for energy saving projects (Deng & Qi 2012, p. 1). According to the European Commission Joint Research Centre on Energy Efficiency (2016, n.p.),

> The approach [EPC] is based on the transfer of technical risks from the client to the ESCO based on performance guarantees given by the ESCO. In EPC, ESCO remuneration is based on demonstrated performance; a measure of performance is the level of energy savings or energy service. EPC is a means to deliver infrastructure improvements to facilities that lack energy engineering skills, manpower or management time, capital funding, understanding of risk, or technology information. Cash-poor, yet creditworthy customers are therefore good potential clients for EPC.

EPC was first introduced to China as a concept in 1997, and therefore had a relatively short development history (Zhang, Li & Chen 2011, p.1377). In China, EPC was also referred to as Energy Management Contracting (EMC) and was regarded as an advanced energy saving model. Companies which provided energy management services were described as "energy saving doctors". This was because they could "diagnose" clients' specific energy problem, provide "prescription" and "treatment" methods, and eventually cure the energy inefficiency (NDRC Foreign Capital Department Yahang Technology Support ECER Research Group 2009, p. 56).

EPC in China had gone through three development stages which roughly paralleled with the 10th (2001–2005), 11th (2006–2010), and 12th (2011–2015) Five-Year Plans. In the initial stage, the Chinese government focused on introducing the concept of EPC and conducted a range of preparation work such as bringing in international funds and setting up demonstration ESCOs. In June 2000, the former State Economic and Trade Commission announced the *Notification for the Further Promotion of EPC*, which was the first official document announced by a central government ministry on the promotion of EPC (EMCA 2010, n.p.). In 2003, a national association for the ESCO industry was established, namely the ESCO Committee of China Energy Conservation Association (EMCA). EMCA received support from NDRC, Ministry of Finance (MF) the World Bank (WB), and Global Environment Facility (GEF), and had a mission of promoting EPC, and facilitating the sustainable, fast, and healthy development of Chinese ESCOs (EMCA 2016, n.p.). Following this, three ESCO pilots were established including Beijing Yuanshen Energy Saving Technology Co., Ltd.,

Liaoning Energy Saving Technology Development Co., and Shandong Energy Saving Engineering Co., Ltd. The three ESCOs received a total of US$63,000,000 low-interest loan from the WB as circulating fund and a combined US$27,000,000 donation from the European Commission and GEF. By the end of 2006, the three demonstration ESCOs completed 450 EPC projects with total investment of over 600,000,000 yuan (approximately US$100,000,000) (Deng & Qi 2012, p. 2).

In the second developmental stage, the government focused on identifying the targeting fields of EPC and making EPC a formalised and systematic industry. For instance, on 14 February 2006, NDRC, Government Offices Administration of the State Council, MF, Central Administration of Management, and PLA General Logistics Department jointly announced the *Notification on the Reinforcement of Resource Saving within Government Organisations*, which clearly required government organisations to employ professional ESCOs to conduct energy saving renovation and optimised energy management in order to achieve energy efficiency improvement (The Central People's Government of PRC 2006b, n.p.). On 4 July 2006, the Office of the Ministry of Construction (MC) released the *Outline of City Green Lighting Program during the 11th Five-Year Plan*, which raised to utilise EPC in undertaking city lighting renovations. It encouraged local governments to employ professional ESCOs in performing a range of comprehensive services such as project design, procurement, construction, training, operation, maintenance, energy efficiency monitoring, and auditing for infrastructures (Office of MC 2006, n.p.). The central government also paid much attention to providing opportunities for ESCOs to grow and improve their services. On 7 April 2006, NDRC and several other departments jointly released the *Notification about the Implementation Plans of a Thousand Enterprises' Energy Saving Activities*. The *Notification* proposed to establish a professional energy saving service system through EPC. The ESCOs in this category would provide one-stop service for the one thousand enterprises including energy inefficiency diagnosis, financing, design, construction, project operation, and management (NDRC 2006, n.p.). The numbers of Chinese ESCOs increased relatively fast from 76 in 2006 to 782 in 2010 because of the supporting policies (Wang 2011, n.p.). However, within this stage, the central government did not provide cash rewards for ESCOs' energy saving achievements and only required local government to provide moderate prizes for outstanding ESCOs as encouragement (MC 2005, p. 55).

The Chinese government started to actively promote EPC and identified EPC as a very important channel for ECERS during the 12th Five-Year Plan, which was regarded as the third developmental stage of EPC in China. Several important EPC policies in regards to financial rewards and tax benefits were announced and implemented, which had a significant impact on the development of China's ESCO industry.

On 2 April 2010, the State Council approved *NDRC and Fellow Departments' Recommendation on Speeding up EPC and Suggestions in Promoting ESCO Industry*

Development, in which cash rewards and tax breaks were proposed for qualified ESCOs. Meanwhile it encouraged banks and other financial institutions to invest in new credit products suitable for ESCOs and to simplify the loan application process (State Council of PRC 2010, n.p.). Three domestic banks, Export-Import Bank of China, Huaxia Bank, and Minsheng Bank were appointed to circulate US$400,000,000 worth of international funds in order to better serve Chinese ESCOs' financing needs (Liu YY 2011, n.p.). The *Recommendation* also set up China's EPC developmental targets by stages. It suggested, by 2012, a range of professional ESCOs with specialty should be established, among which large and comprehensive ESCOs should be in proportion. A healthy, orderly, and lively energy service market should also be established by then. By 2015, EPC should become one of the major procedures for energy end users' energy saving upgrades and renovations. A relatively complete energy saving service system should be established, where ESCOs become stronger and capable of providing broader services in broader areas. (State Council of PRC 2010, n.p.).

On 3 June 2010, MF and NDRC jointly announced the *Temporary Management Method for the Financial Reward Fund of EPC Projects*, which stated the detailed amount of cash rewards for qualified ESCOs. To be qualified, an ESCO's contracted project needed to achieve an annual energy saving between 100 tce and 10,000 tce (Industrial projects needed to be between 500 tce and 10,000 tce). For qualified projects, the central government would reward the responsible ESCO 240 yuan/tce (approximately US$40), and local government would provide no less than 60 yuan/tce (approximately US$10) to the same ESCO. NDRC and MF would conduct continuous monitoring and management for qualified ESCOs and made the list of qualified ESCOs and their business scopes public (MF & NDRC 2010, n.p.). This policy largely accelerated the development of the Chinese ESCOs industry, however, it also planted hidden burdens similar to that of the energy saving and emission reduction cash rewards directly distributed to enterprises. Fraudulent, unlawful, and exaggerated claims was a big problem during the implementation, and this will be discussed later in this chapter.

Apart from cash rewards, MF and State Taxation Administration (STA) announced the *Notification Value Added Tax, Business Income Tax, and Enterprise Income Tax Policies for Promoting ESCO Industry* on 30 December 2010. This policy temporarily exempted business tax and value added tax for qualified ESCOs; it also exempted the first three year's business income tax (counting from the first business income gain) for qualified ESCOs and charged only half of the statutory tax rate from the fourth year to the sixth year. To be qualified for the tax benefit, the technology and the contract of an EPC project should be strictly in accordance with the *General Technical Rules for EPC* released by State General Administration of PRC for Quality Supervision, Inspection and Quarantine (AQSIQ) and Standardisation Administration of PRC (SAC). This policy was implemented from 1 January 2011 (MF & STA 2010, n.p.).

Nonetheless, EPC, as a relatively new phenomenon, experienced high-speed development during the 11th and 12th Five-Year Plan periods (2006–2015). According to EMCA statistics, China's EPC investment increased from 1,300,000,000 yuan (approximately US$217,000,000) in 2005 to 55,700,000,000 yuan (approximately US$9,300,000,000) in 2012, and the total output of EPC industry increased from 4,700,000,000 yuan (approximately US$80,000,000) to 165,300,000,000 yuan (approximately US$27,500,000,000) during the same period which led to a 34-fold growth of the EPM market. By the end of 2012, there were a total of 4175 Chinese ESCOs nationwide, whose EPC projects realised a total of 18,280,000 tce energy saving and reduced 45,710,000 tons of CO_2 emission (China Environmental Protection Online 2013, n.p.).

Successful examples of enterprises utilising EPC for energy saving

The fast development of EPC proved to be an effective mechanism and brought much benefit to enterprises which utilised this service in their energy saving activities, especially the medium to small enterprises which lacked technological capacity and capital but had great potential for energy savings. Many successful examples had reflected the advantages of this new mechanism.

Case 1: EPC between Jiaxing Shengdi Lighting Technology Co., Ltd. and Yicheng Textile Co., Ltd.

In 2013, Yicheng Textile Co., Ltd. of Xinteng township completed its lighting system renovation under an EPC agreement with Jiaxing Shengdi Lighting Technology Co., Ltd. By completion, 1250 sets of fluorescent lights in the workshop were replaced with LED lights. The new lighting system would save over 60% of electricity compared with the old system. The good thing about EPC was that Yicheng Textile Co., Ltd. did not have to pay anything upfront for the renovation; and Jiaxing Shengdi Lighting Technology Co., Ltd. would recoup its investment and extract profit from Yicheng Textile's energy saving fund within three years.

According to the manager of Yicheng Textile, lighting used to cost the company about 150,000 yuan (approximately US$25,000) a year. The fluorescent light bulbs also needed to be changed two to three times a year which increased the cost. The LED lights that Shengdi Lighting provided would use 60% less energy and would last over 25,000 hours, which meant big energy saving and significantly smaller electricity bills for Yicheng Textile. According to the contract between the two companies, Shengdi Lighting would pay for the entire cost of the renovation of 160,000 yuan (approximately US$28,000) upfront; and Yicheng Textile would repay Shengdi

Lighting's entire project costs within three years from its energy saving with the first year 90% of its saving, the second year 80% of its saving, and the third year 50% of its saving. By the end of the three years, all the LED lighting would belong to Yicheng Textile. These two companies created a win-win solution through EPC. More importantly, Yicheng Textile's annual energy saving meant they avoided burning 42.5 tons of standard coal each year which brought positive CO_2 emission reduction and contribution to global climate stability (Alighting 2013, n.p.).

> ***Case 2: EPC among Shanxi Qinling Cement (Group) Co., Ltd., Siemens IA&DT, and a Shanxi ESCO***
>
> Shanxi Qinling Cement (Group) Co., Ltd. (Qinling Cement) has a history of over half a century and is one of China's top 60 significant cement enterprises. Cement is an important material for China's infrastructure and urbanisation; however, it consumes a large amount of electricity. Qinling Cement faced a great challenge in technological upgrade in achieving energy saving and emission reduction. Siemens spotted that China's Cement industry faced more urgent needs for energy saving compared with other industries; therefore, it led the renovation project of replacing Qinling Cement's fixed-frequency kiln fans through EPC with a local ESCO.

The old kiln fans at Qinling Cement consumed 30–40% of the factory's total electricity and were designed to run at full capacity at all times. Upgrading to Siemens' frequency conversion fans meant a significant amount of energy saving. Siemens assisted Qinling Cement in conducting detailed technological and feasibility analyses prior the EPC, and the contract was signed successfully at the end of 2007 with the involvement of a local ESCO in Shanxi. The ESCO spent all the upfront cost in purchasing the Siemens frequency conversion fans and loaned them to Qinling Cement. During the contract term, Qinling Cement would repay the ESCO with part of the savings from its reduced energy bills. By the completion of the EPC, Qinling Cement were entitled to keep the fans, the ESCO recouped its investment and made profit, and Siemens increased its sales. This was a triple win scenario as Siemens and the ESCO both made a profit and Qinling Cement would continue saving money from new fans and reducing energy consumption and CO_2 emission (Zhang & Yin 2009, pp. 50–52).

The above examples indeed reflected the advantages of EPC and proved the feasibility of EPC in China's ECERS. With support from the central government and the diversification of funding sources, EPC gradually got known and recognised. However, there are some factors that restricted the large-scale expansion of EPC in China. In the next section, these factors will be discussed.

Factors that restrict the flourishing of EPC in China

EPC, as an effective market mechanism for energy efficiency management, has spread across multiple sectors since its introduction to China by the WB in 1996. EPC projects covered China's industrial sector; heating, cooling, and lighting of the building sector; and the transport sector. However, according to Tang and Zou (2013, n.p.), EPC should have expanded to a larger scale than it had since the government utilised EPC as one of the important market tools to achieve the ECERS results. However, there were some obstacles that restricted the flourish of China's EPC industry, namely ESCOs' financing difficulty, the underdeveloped risk analysis system, and difficulty in finding suitable projects.

Financing difficulty was widely experienced by many ESCOs, and especially small ESCOs. According to the data collected by Ju U Neng (a service platform for energy efficiency monitoring and data collection), over 70% Chinese ESCOs faced some sort of financing challenge (Ju U Neng 2015, n.p.). One of the characteristics of EPC was that in the majority of cases ESCOs were responsible for upfront financial input and only harvested profit after the projects' completion from their clients' financial gains from energy saving. As a result, it was a long-term process for ESCOs to recoup their total investment and gain profit, which usually lasted several years depending upon individual contracts. If the contract term was especially long, it might put financial stress on ESCOs. Trying to solve the financial difficulty through banks was not easy either, as the majority of Chinese ESCOs were medium to small in scale, and therefore lacked of assets such as land and real estate which banks required for loans against collateral (China Environmental Protection Online 2013, n.p.).

Another risk that ESCOs faced was whether or not they could fully recoup their investment. The nature of EPC is that an ESCO and an energy end user realise benefit sharing through a contract; therefore, the possibility of breaking a contract exists. Hence, an ESCO is required to be extra cautious when selecting clients. There is risk associated with the credit of the energy end user. If the energy end user experiences bankruptcy before the term, an ESCO will be unable to recoup its total investment and will bear financial loss. Moreover, the majority of EPC terms are quite long, which amplifies this potential risk. In addition, an ESCO also bears a range of risks such as financial risks, design and technological risks, and the quality risks of equipment and materials passed down from its suppliers. Any of these risks may put an ESCO into a difficult position and cause dispute with its clients (Song, Zeng & Long 2011, p. 36).

Although selecting reliable clients can reduce ENCOs' risks, this is not an easy task. This is because most large enterprises with reliable credit already have the financial and technological capacity in conducting energy saving on their own; while many small enterprises that need EPC have problems such as hard-to-predict energy consumption and unstable energy saving during the fluctuation of the market and their operation, which makes it risky for ESCO's investment return (Ju U Neng 2015, n.p.).

Within the analysed period, a favourable credit environment and atmosphere for EPC was lacking. The lack of trust became the biggest restriction for EPC development. Some enterprises worried that ESCOs' technology could be unreliable; while ESCOs were concerned that enterprises might find excuses for delaying repayments after the EPC was implemented (Wang 2008, p. 22). In order to target this issue, risks for both ESCOs and energy saving enterprises should be managed so that trust can be built up between the two bodies. This will be discussed in Chapter 6.

Apart from the above challenges, some ESCOs themselves posed obstacles for the healthy development of the Chinese EPC industry. According to NDRC's review on the 2010 EPC financial reward claims handed in by all localities, the exaggeration of energy saving amounts was one of the biggest problems. There were also other problems such as the amount of energy saving was not within the reward category, the technology used was not supported by the cash reward scheme, rewards took up too big a proportion of the entire investment, or the contract was signed long before the policy was implemented. In regards to these problems, NDRC and MF jointly announced the *Notification on Further Reinforcing the Supervision and Inspection of EPC Projects* on 20 July 2011. The *Notification* required local energy saving departments to liaise with local finance departments in conducting immediate onsite inspections of EPC projects within their administrative regions. During the inspection, the following aspects of EPC projects should be checked strictly according to MF's relevant documents and criteria: the authenticity, the total energy saving amount, technical and economic index, the time of contract signed, and the detailed content and procedures of the EPC projects. Local energy saving departments and finance departments should be responsible for the inspection, while ESCOs and third-party auditing organisations should be responsible for the authenticity of the information they provided. ESCOs were required to return all cash rewards if they were found providing false information or cheating. ESCOs were required to return partial cash rewards if they were found exaggerating their energy saving amounts. Any ESCOs and third-party auditing organisations which submitted fraudulent claims would be disqualified from business (NDRC Office & MF Office 2011, n.p.).

The financial reward scheme for qualified ESCOs was implemented through the majority of the 12th Five-Year Plan. The unlawful claims from many ESCOs shook the government's decision on the continuation of ESCO financial reward policy. On 2 May 2015, MF announced the abolition of cash rewards to ESCOs together with the abolition of cash rewards for phasing out backward production capacities in the new *Temporary Management Method for Energy Saving and Emission Reduction Financial Reward Fund [Finance & Construction 2015 Document No. 161]*. Since then, qualified ESCOs only enjoyed tax benefit but no more cash rewards (MF 2015, n.p.).

The development of EPC in China was relatively fast but not yet on the same scale as Western developed countries. From China's evolving EPC policies we could spot that EPC faced similar problems with other ECERS programs. For instance, fraudulent claims of government cash grants could not be eliminated through ESCOs. As a result, the government's financial input could not achieve full capacity in energy saving and emission reduction. That was why the government decided to abolish the general cash rewards scheme for the 13th Five-Year Plan and only provided funding for specific projects within the government's specific targeted areas (MF 2015, n.p.). In order to address the cost-effectiveness of ECERS, the Chinese government should pay more attention to new mechanisms that can avoid or at least reduce fraudulent claims and corruption. Discussion of one possibility of such an innovation will be presented in Chapter 6.

The effectiveness of ECERS

The command and control implementation procedures in ECERS could be described as effective overall; however, they were considerably more expensive than policies utilising market-based instruments. According to Qi (2013a, p. 234), China's command and control style domestic ECERS was 10 times as costly as CDM projects for each unit of CO_2 reduction. In developing countries, CDM costed from several US dollars up to 20 US dollars for each ton of CO_2 reduction, while in China's domestic programs, it costed an average of 1416 yuan (approximately US$220) during the 11th Five-Year Plan. It was obvious that the efficiency of China's domestic ECERS did not match the international market mechanism, CDM. The Chinese government also realised the costly nature of administrative measures and actively experimented with market mechanisms such as CCER and EPC. Although these two market mechanisms are not yet mature in China, it deserves thorough investigation of whether or not a combination of the two mechanisms could bring some effectiveness and efficiency improvement to ECERS. This will be explored in Chapter 6.

Summary

Overall, ECERS contributed to China's energy structure optimisation and climate change mitigation positively during the 11th and 12th Five-Year Plans. It solved tough problems such as the closure of small inefficient thermal plants within a short period of time. This was a powerful aspect of China's administrative system; the policies in this area also appeared to be innovative and effective. However, there were cases of inadequate compensation and unfair treatment of laid-off workers which need to be addressed. There were also a range of weaknesses exposed during the implementation of ECERS. Fraudulent claims of cash rewards for unachieved energy saving and emission reduction was a serious problem. It not only wasted financial resources, but also shook the accountability of ECERS.

Small- and medium-scale renewable energy companies experienced considerable challenges compared with those under big energy SOEs. There were also cases of provincial and local governments who seriously violated the renewable energy law by giving priority to coal power, thus restricting power generated from renewable energy sources. The unfair competition and market environment was detrimental for the long-term sustainability of China's renewable energy development. In order to improve the transparency and efficiency of ECERS, the Chinese government introduced the third-party auditing organisations and the market mechanism EPC; however, they could not fundamentally eliminate false reporting and fraudulent claims of cash rewards. These problems seriously damaged the overall accountability of ECERS and pushed up the total cost of China's domestic energy saving and emission reduction, and therefore, deserve in-depth study and careful consideration. In the next chapter, provided are some recommendations for the problems mentioned in this chapter along with those raised in Chapters 3 and 4.

List of Interviewees

Interviewee 1, Officer of NDRC

Interviewee 7, Professor at the Centre for Rural Environmental and Social Research, CASS

Interviewee 8, Officer of National Energy Administration, NDRC

Interviewee 9, Previous private owner of a small coal-fired power plant

Interviewee 10, Head of an ENGO (Organisation name wished not to be identified)

Interviewee 11, Owner of a private small-scale wind farm

Interviewee 12, Consultant of Saidi Zhiku Consultancy

Interviewee 13, Analyst specialised in renewable energy

Interviewee 14, Professor from the School of International Studies, CASS

Note

1. Two sessions: the National People's Congress (NPC) and the National Committee of the Chinese People's Political Consultative Conference (CPPCC).

References

Alighting 2013, *Hetong nengyuan guanli shixian jieneng jianghao shuangying (EPC realised win-win solution in energy saving and emission reduction)*, viewed 02 July 2015, <http://www.alighting.cn/news/2013625/n663853115.htm>.

Baidutieba 2016, *Mengxi dianli xitong you duoshao xiagang zhigong Debussy tuoshan anzhi? (How many laid-off workers did not receive appropriate arrangements in the Inner Mongolia West Electricity System?)*, viewed March 2016, <http://tieba.baidu.com/p/4303408279?pn=2>.

BBS163 2014, *Jiaolv, xiaohuodian guanbi, Shui zaihu xiagang gongren? (Frustration, small coal-fired power plant closed, who cares about the laid-off workers?)*, viewed 29 July 2014, <http://bbs.travel.163.com/bbs/tuyou/185190528.html>.

Cai, W, Wang, C, Chen, J & Wang, S 2011, 'Green economy and green jobs: myth or reality? The case of China's power generation sector', *Energy*, vol. 36, no. 10, pp. 5994–6003.

Cao, HC 2014, *Huodianchang kuaojian wuran kong qi? (Coal-fired power plant expansion cause air pollution?)*, viewed 22 March 2016, <http://jl.cepnews.com.cn/news_348887.html>.

CCTV News 2015, *Guojia nengyuanju "shierwu" jiahao mubiao chaoe wancheng: zhongguo chengwei shijieshang xinnengyuan liyong diyi daguo (National Energy Bureau stated that energy consumption reduction exceeded target during the 12th Five-Year Plan: China became the biggest renewable energy utilisation nation)*, viewed 08 April 2016, <http://news.cntv.cn/2015/12/30/VIDE1451424844267390.shtml>.

CEC 2011, *Dianli hangye shierwu guihua - youhua fazhan meidian (Power Industry 12th Five-Year Plan - Optimising development of coal power)*, viewed 02 June 2016, <http://www.cec.org.cn/zhuanti/2011nianzhuantihuigu/dianlixingyeshierwuguihua/youhuadianyuanjiegouhebuju/2011-01-18/38643.html>.

Chen, S, Sun, Z, Tang, S & Wu, D 2011, 'Government intervention and investment efficiency: evidence from China', *Journal of Corporate Finance*, vol. 17, no. 2, pp. 259–271.

Chen, YY 2011, *40 jia qiye nuoyong taoqu jieneng jianpai zijin bei dianming (40 enterprises fraudulently claimed ECERS funds being identified)*, viewed 22 June 2015, <http://www.cs.com.cn/xwzx/05/201105/t20110514_2878898.html>.

Cheng, YS 2009, 'Xiaohuodian jizu guanting zhongde wenti fenxi he duice tantao (Problems emerged during the closure of small coal-fired power plants and solutions)', *Technology Economic and Market*, vol. 5, pp. 108–109.

China Daily 2015, *Full text of President Xi's speech at opening ceremony of Paris Climate Summit*, viewed 08 April 2016, <http://www.chinadaily.com.cn/world/XiattendsParisclimateconference/2015-12/01/content_22592469.htm>.

China Electricity Council 2011, *Dianli hangye quanmian wancheng "shiyiwu" jieneng jianpai zhibiao (The power sector completed ECERS targets of the 11th Five-Year Plan)*, viewed 08 January 2016, <http://www.cec.org.cn/yaowenkuaidi/2011-10-25/72654.html>.

China Electricity News Network 2007, *Guowuyuan yaoqiu xianqi shangbao guanting xiaohuodian fang'an (The State Council set deadline for submitting plans on the closure of small coal-fired power plants)*, viewed 02 February 2016, <http://finance.sina.com.cn/chanjing/b/20070227/09331232380.shtml>.

China Environmental Protection Online 2013, *Guojia dali fuchi hetong nengyuan guanli shichang huode changzu fazhan (EPC market gained long-term development due to great government support)*, viewed 15 August 2016, <http://www.hbzhan.com/news/detail/81123.html>.

Cong, J & Guo, JF 2012, *Kongke gongchang taoqu qianwan jieneng jianpai buzhu zijin anfan quan huoxing (Empty shell factory cheated for ten million energy conservation and emission reduction support fund, criminals are under custody)*, viewed 27 May 2015, <http://news.sohu.com/20120406/n339937491.shtml>.

Deng, XH & Qi, Y 2012, 'Sinicization and development status of energy performance contracting', *Environmental Science and Management*, vol. 37, no. 12, pp. 1–6.

EMCA 2010, *Guojia jingmaowei ziyuan jieyu yu zonghe liyong si–guanyu jinyibu tuiguang "hetong nengyuan guanli" jizhi de tonggao (State Economic and Trade Commission Department of Resource saving and Comprehensive Utilisation announced The Notification for the Further Promotion of EPC)*, viewed 15 August 2016, <http://policy.emca.cn/n/20130306032625.html>.

EMCA 2016, *Brief introduction of EMCA*, viewed 15 August 2016, <http://about.emca.cn/>.

European Commission 2012, *Questions & answers on use of international credits in the third trading phase of the EU ETS (January 2012)*, viewed 19 April 2016, <http://ec.europa.eu/clima/policies/ets/linking/faq_en.htm>.

European Commission Joint Research Centre on Energy Efficiency 2016, *Energy Performance Contracting*, viewed 10 August 2016, <http://iet.jrc.ec.europa.eu/energyefficiency/european-energy-service-companies/energy-performance-contracting>.

EWEA 2005, *Large scale integration of wind energy in the European power supply: analysis, issues and recommendations*, viewed 15 July 2016, <http://www.uwig.org/eweastudy/051215_grid_report.pdf>.

Fenghuang Finance 2015, *Fengdianye zaixian kongbu yimu qifengxiandian zai shengji (Wind power industry again showed venerability - wastage and restriction of wind power upgraded)*, viewed 16 July 2016, <http://finance.ifeng.com/a/20151110/14063693_0.shtml>.

Ferris Jr., RJ & Zhang, HJ 2015, 'Environmental law in the People's Republic of China', in KA Day (ed), *China's environment and challenges of sustainable development*, Routledge, London & New York, pp. 66–101.

Guo, G 2007, 'Retrospective economic accountability under authoritarianism evidence from China', *Political Research Quarterly*, vol. 60, no. 3, pp. 378–390.

He, X 2012, *Shandong Xingquan jituan taoqu jigai zijin 800wan, pianqu jiangli 245wan (Shandong Xingquan Group cheated for 8 million yuan technology upgrade grant and fraudulently claimed 2.45 million yuan energy conservation and emission reduction cash reward)*, viewed 27 May 2015, <http://www.ce.cn/XWZX/fazhi/201208/07/t20120807_23563777.shtml>.

Hu, XQ 2011, *Fagaiwei guwen: fangzhi gebie qiye yi jieneng xiangmu pian jiangli, "disanfang" muqian shuliang pianshao (NDRC consultant: preventing individual enterprises' cheating behaviour for energy saving cash reward, the number of third party auditing bodies currently too few)*, viewed 07 May 2015, <http://news.xinhuanet.com/energy/2011-11/c_122348993_2.htm>.

Hua, TL 2015, *Redianchang meichen "ranhei" gandoufu xu (Coal dust from coal-fired power plant dyed black dried bean curd: TSP density exceeded the standard by 2.31 times)*, viewed 07 April, 2016, <http://www.jl.xinhuanet.com/2012jlpd/2015-03/17/c_1114660671.htm>.

iChina Energy 2016, *Qin Haiyan: jiejue qifeng "wanji" yao "dongzhendao" (Qin Haiyan: to solve the wastage of wind energy problem requires tough penalty)*, viewed 11 July 2016, <http://www.ichinaenergy.com/News/1/96543.html>.

Ju U Neng 2015, *Jiyu shuangying jichu de hetong nengyuan guanli weihe chichi zuobuda? (Why is the win-win EPC not expanding into large scale?)*, viewed 17 August 2016, <http://www.sdeex.com/Infoservice/info.aspx?id=21172>.

Lantian baiyun zhongkeqi 2016, *Wumai zhenxiang – yige huanbao bumen gongwuyuan de jishouzibai (The truth of city smog - confessions of an EPB public servant)*, viewed 7 July 2017, <http://www.lvngo.com/ngo-43381-1.html>.

Li, F, Dong, SC, Li, X, Liang, QX & Yang, WZ 2011, 'Energy consumption-economic growth relationship and carbon dioxide emissions in China', *Energy Policy*, vol. 39, no. 2, pp. 568–574.

Li, XW 2007, 'Guanting xiaohuodian sixiang zhuyi (Four issues of closure of small coal-fired power plants)', *CPNN*, viewed 25 February 2016, <http://www.cpnn.com.cn/jdwt3/gd/200704100047.htm>.

Lin, J 2007, 'Energy conservation investments: a comparison between China and the US', *Energy Policy*, vol. 35, no. 2, pp. 916–924.

Liu, C, Kong, XL & Gao, TM 2008, 'Empical analysis of change in China's industrial sector energy consumption intensity and influential factors', *Resources Science*, vol. 9, no. 30, pp. 1290–1299.

Liu, GY 2011, 'Guanghui wunian – "shiyiwu" dianli gongye fazhan chengjiu feiran (Glorious five years – the achievements of the electricity sector in the 11th Five-Year Plan)', *China Power Enterprise Management*, vol. 1, pp. 8–13.

Liu, RF, Liu, WG, Zhang, W, Liu, J & Chen TE 2014, 'Economic evaluation of the trans-provincial and trans-regional renewable energy trading based on the renewable energy quota system', *Grid and Renewable Energy*, vol. 30, no. 1, pp. 59–63.

Liu, Y & Kokko, A 2010, 'Wind power in China: policy and development challenges', *Energy Policy*, vol. 38, no. 10, pp. 5520–5529.

Liu, YY 2011, *Hetong nengyuan guanli: fengyu shiwunian daoyuan ren gengzhong (Energy Management Contracting: 15 years of development and still a long way to go)*, viewed 11 August 2016, <http://www.chinajsb.cn/bz/content/2011-01/24/content_19556.htm>.

MC 2005, 'Guanyu jiaqiang chengshi zhaoming guanli cujin jieyue yongdian gongzuo de yijian (Suggestions on improving the managing of city lighting and saving electricity)', *Deng yu Zhaoming*, vol. 29, no. 1, pp. 55–56.

MF & NDRC 2010, *Guanyu yinfa hetong nengyuan guanli xiangmu caizheng jiangli zijin guanli zanxing banfa - CaiJian [2010] 249 hao (The Temporary Management Method for the Financial Reward Fund of EPC Projects - CaiJian [2010] Document No.249)*, viewed 13 July 2016, <http://www.gov.cn/zwgk/2010-06/09/content_1623756.htm>.

MF & STA 2010, *Guanyu cujin jieneng fuwu chanye fazhan zengzhishui yingyeshui he qiye suodeshui zhengce wenti de tongzhi - Caizheng [2010] 110 hao (Notification Value Added Tax, Business Tax and Enterprise Income Tax Policies for Promoting ESCO Industry - Finance & Tax [2010] Document No. 110)*, viewed 8 July 2016, <http://www.gov.cn/zwgk/2011-02/17/content_1804866.htm>.

MF 2011, *Guanyu yinfa <jieneng jishu gaizao caizheng jiangli zijin guanli banfa> de tongzhi – Caijian [2011]367 hao (Notification on the < The Management Method for Energy Saving Technological Upgrade Financial Reward Fund> Finance and Construction [2011])*, viewed 20 December 2015, <http://www.mof.gov.cn/pub/jinjijianshesi/zhengwuxinxi/zhengcefagui/201106/t20110623_566086.html>.

MF 2015, *Guanyu yinfa jieneng jianpai buzhu zijin guanli zanxing banfa de tongzhi - Caijian [2015] 161 hao (Notification on the release of Temporary Management Method for Energy Saving and Emission Reduction Financial Reward Fund - Finance & Construction [2015] Document No. 161)*, viewed 20 August 2016, <http://jjs.mof.gov.cn/zhengwuxinxi/zhengcefagui/201505/t20150519_1233458.html>.

NBSC 2016, *Niandu shuju - 2015 Ningxia Fengdian (Annual data - Nixia wind power)*, viewed 01 June 2016, <http://data.stats.gov.cn/search.htm?s=宁夏风电>.

NDRC 2006, *Guanyu yinfa qianjia qiye jieneng xingdong shishi fang'an de tongzhi - Fagai Huanzi [2006] 571hao (The Notification about the Implementation Plans of A Thousand Enterprises' Energy Saving Activities - NDRC Environment and Resource Department [2006] Document No. 571)*, viewed 18 July 2016, <http://www.sdpc.gov.cn/rdzt/jsjyxsh/200604/t20060413_66114.html>.

NDRC Foreign Capital Department Yahang Technology Support ECER Research Group 2009, 'Zhongguo hetong nengyuan guanli chanye fazhan zhuangkuang he jianyi (The developmental status and recommendations of Chinese EFC industry)', *China Technology Investment*, vol. 8, pp. 56–58.

NDRC Office & MF Office 2011, *Guanyu jinyibu jiaqiang hetong nengyuan guanli xiangmu jiandu jiancha gongzuo de tongzhi - Fagaiban Huanzi [2011] 1755 hao (Notification on Further*

Reinforcing the Supervision and Inspection of EPC Projects - NDRC Administrative Office Environment and Resources [2011] Document No. 1755), viewed 13 July 2016, <http://www.sdpc.gov.cn/zcfb/zcfbtz/201107/t20110726_424836.html>.

NEA 2012, *Minjian touzi zai xinnengyuan he kezaisheng nengyuan lingyu fahui le zhongyao zuoyong (Private investment played an important role in new and renewable energy industry)*, viewed 09 April 2015, <http://www.nea.gov.cn/2012-05/18/c_131595506.htm>.

NEA 2016, *Guojia nengyuanju guanyu zuohao 2016 niandu fengdian xiaona gongzuo youguan yaoqiu de tongzhi, GuoNeng XinNeng [2016] 74 hao (NEA notification on the requests of managing wind power grid connection and utilisation in 2016, NEA Renewable Energy [2016] No. 74 Document)*, viewed 20 May 2016, <http://www.gov.cn/xinwen/2016-03/18/content_5055168.htm>.

Office of MC 2006, *Guanyu yinfa "shiyiwu chengshi lvse zhaoming gongcheng guihua gangyao de tongzhi - Jianbancheng [2006] 48 hao (The Outline of City Green Lighting Program during the 11th Five-Year Plan - MC [2006] Document No. 48)*, viewed 12 July 2016, <http://www.gov.cn/zwgk/2006-07/10/content_331957.htm>.

Qi, Y & Wu, T 2013, 'The politics of climate change in China', *Wiley Interdisciplinary Reviews: Climate Change*, vol. 4, no. 4, pp. 301–313.

Qi, Y 2013, *Annual review of low-carbon development in China: 2010*, World Scientific Publishing, Singapore.

Qin, HY 2016, *Qin Haiyan: Gansusheng qifeng xiandian kunju jiexi (Qin Haiyan: An analysis of Gansu Province's dilemma in abandoning wind power)*, viewed 20 June 2016, <http://news.xinhuanet.com/energy/2016-04/12/c_128887670.htm>.

Shenzhen City News 2015, *Heyuan dianchang wuran yingxiang jihe? (What is the pollution impact of Heyuan Coal-fired Power Plant?)*, viewed 28 March 2016, <http://dtzbd.sznews.com/html/2015-04/15/content_3196852.htm>.

Shi, D 2012, 'Woguo xin nengyuan channeng "guosheng" de yuanyin yu jiejue tujing (The reasons and solution to China's overcapacity in the new and renewable energies)', *China Energy*, vol. 9, pp. 20–23, viewed 19 April 2016, <http://www.cnki.com.cn/Article/CJFDTotal-ZGLN201209003.htm>.

Song, YQ, Zeng, Y & Long, WD 2011, 'The development and analysis of energy performance contracting in China', *Building Energy and Environment*, vol. 30, no. 2, pp. 33–36.

South China Morning Post 2016, *China wasted enough renewable energy to power Beijing for an entire year, says Greenpeace*, viewed 10 August 2017, <http://www.scmp.com/news/china/policies-politics/article/2088865/china-wasted-enough-renewable-energy-power-beijing>.

Stanway, D 2014, 'China approves massive new coal capacity despite pollution fears', *Reuters*, viewed 5 May 2016, <http://www.reuters.com/article/china-coal-idUSL3N0K90H720140107>.

State Council of PRC 2006, *'Guowuyuan guanyu "shiyiwu" qijian gediqu danwei shengchan zongzhi nengyuan xiaohao jiangdi biaozhun de pifu - guohan [2006] 94 hao (State Council approved per unit GDP energy consumption reduction targets for all districts - [2006 State Letter No. 94])'*, viewed 8 August 2016, <http://www.gov.cn/gongbao/content/2006/content_443285.htm>.

State Council of PRC 2007, *Guowuyuan bizhuan fazhan gaige wei nengyuanban guanyu jiakuai guanting xiaohuodian jizu ruogan yijian de tong zhi - guofa [2007] 02 hao (State Council approved Several Suggestions on the Acceleration of Closing Down Small Thermal Power Plants issued by NDRC and the State Energy Office - [2007 State Document No. 02])*, viewed 1 February 2015, <http://www.gov.cn/gongbao/content/2007/content_534198.htm>.

State Council of PRC 2010, *Guowuyuan bangongting zhuanfa fazhan gaige wei de bumen guanyu jiakuai tuixing hetong nengyuan guanli cujin jieneng fuwu chanye fazhan yijian de tongzhi - guobanfa [2010] 25 hao (The State Council forwarding NDRC and fellow departments' notification on speeding up EPC and suggestions in promoting ESCO industry development [2010 State Council Office Document no. 25])*, viewed 15 August 2016, <http://www.gov.cn/zwgk/2010-04/06/content_1573706.htm>.

State Grid Corporation of China 2012, *"Shiyiwu" shiqi woguo nengyuan fazhan gaikuang (General information of China's energy development in the 11th Five-Year Plan)*, viewed 08 January 2016, <http://www.sgcc.com.cn/xwzx/nyzx/2012/06/275508.shtml>.

Sun, GB, Zhu, C, Ruan, YG & Wang, LM 2008, 'Xiaohuodian jizu guantinghou mianlinde wenti jie chuli cuoshi' (Problems after the closure of small coal-fired power plants and solutions)', *Anhui Power Technology Information*, vol. 2, pp. 30–35.

Sun, Q 2009, Deputy Director of NEB, *"Shiyiwu" guanting xiaohuodian jizu renwu tiqian chao'e wancheng xinwen fabuhui (Sun Qing's response to journalists, the closure of small coal-fired power plants during the 11th Five-Year exceeded target and ahead of schedule)*, viewed 16 February 2016, <http://webcast.china.com.cn/webcast/created/4066/152_1_0101_desc.htm>.

Tang, R & Zou, CL 2013, *Nengyuan hetong guanli "jiaohao bu jiaozuo" shuangying moshi zao lengyu, chengbai guanjian zai fengxian kongzhi (EPC received well but were not largely expanded, the key for success is risk control)*, viewed 17 August 2016, <http://www.cpnn.com.cn/zdyw/201305/t20130514_575517.html>.

The Central People's Government of PRC 2010, *Liang bumen tan "jiakuai taotai luohou channeng, cujing quanguo jieneng jianpai" (Two departments review on "the acceleration of eliminating backward production capacity and promoting energy conservation and emission reduction)*, viewed 14 March 2016, <http://www.gov.cn/zxft/ft199/wz.htm>.

The Central People's Government of PRC 2006a, *Zhonghua renmin gongheguo guomin jingji he shehui fazhan di shiyige wunian jihua guangyao (The 11th Five-Year Plan for National Economy and Social Development)*, viewed 03 January 2016, <http://www.gov.cn/ztzl/2006-03/16/content_228841.htm>.

The Central People's Government of PRC2006b, *Wu bumen lianhe fa tongzhi yaoqiu jiaqian zhengfu jigou jieyu ziyuan gongzuo (Five ministries jointly released the Notification on the Reinforcement of Resource Saving within Government Organizations)*, viewed 15 July 2016, <http://www.gov.cn/jrzg/2006-03/03/content_216753.htm>.

Wang, Q 2010, 'Effective policies for renewable energy – the example of China's wind power – lessons or China's photovoltaic power', *Renewable and Sustainable Energy Reviews*, vol. 14, no. 2, pp. 701–712.

Wang, SM 2008, 'Hetong nengyuan guanli zai woguo de fazhan he cunzai de wenti (The development and survival of EPC in China)', *China Energy*, vol. 30, no. 2, pp. 21–23.

Wang, XL 2011, *Jieneng fuwu hangye yingjie chanye chuntian (ESCOs embraces the spring of the industry)*, viewed 15 July 2016, <http://paper.people.com.cn/zgnyb/html/2011-01/17/content_725181.htm>.

Wang, Y 2015, *Taotai luohou channeng heli zhiding guanting xiaohuodian shijianbiao (Eliminating backward production capacity by making sensible schedule for closure of small coal-fired power plants)*, viewed 10 February 2016, <http://info.gongchang.com/t/zhuangbei-797363.html>.

Wang, YC 2016, *Dianpan "shierwu" huihuang chengjiu: qingjie nengyuan "fangguang wuxian" (Comment on the achievements of the 12th Five-Year Plan: renewable energy became the focal point)*, viewed 08 April 2016, <http://finance.sina.com.cn/roll/2016-02-24/doc-ifxprupc9917617.shtml>.

Weipu 2007, 'Jiu dianli gongye shangda yaxiao jieneng jianpai da jizhewen'(Interview response of energy conservation and emission reduction in the energy industry in terms of replacing small thermal plants with large ones)', *Resource Inhabitant and Environment*, vol. 38, pp. 16–17, viewed 01 February 2016, <http://www.cqvip.com/read/read.aspx?id=23949053>.

Williams, L 2014, 'China's climate change policies: drivers and actors', *Lowy Institute for International Policy*, viewed 11 March 2013, <https://www.files.ethz.ch/isn/182715/chinas-climate-change-policies.pdf>.

Wu, YP 2008, 'Caizheng fenquan, difang baohu zhuyi yu dianli hangye guanzhi xiaolv (Separation of money and power, local protectionism and the management efficiency of the power industry)', *Guoji Yantaohui*, pp. 326–337.

Xue, LQ & Yang, WS 2011, 'Lun zhongguo zhengce zhixing moshi de tezheng: yi "shiyiwu" qijian chenggong guanting xiaohuodian weili (Characteristics of China's policy implementation models: the example of closure of small coal-fired power plants during the 11th Five-Year Plan)', *Public Management*, vol. 8, no. 4, pp. 1–7.

Zhang, WY & Yin, Z 2009, 'Qinling Shuini hetong nengyuan guanli de chenggong dianfan (Qinling Cement the successful example of EMC)', *Dianqi Shidai*, vol. 7, pp. 50–52.

Zhang, XH, Li, X & Chen, SL 2011, 'Problem and countermeasure of energy performance contracting in China', *Energy Procedia*, vol. 5, pp. 1377–1381.

Zhang, YM 2012, *Nengyuan chanye minying qiye reng mian nanju - gongping jingzheng huanjing daiyingzao (Private enterprises facing difficulty in the energy industry - a fair competition environment is awaiting to be created)*, viewed 27 April 2016, <http://roll.sohu.com/20120910/n352734981.shtml>.

Zhou, M 2010, 'Xiao huodian guanting fangshi qianxi (Analysis of the closure of small coal-fired power plants)', *Qiye Daobao*, vol. 12, pp. 270–271, viewed 05 March 2016, <http://www.cqvip.com/read/read.aspx?id=36219140>.

Zhou, N, Levine, MD & Price, L 2010, 'Overview of current energy-efficiency policies in China', *Energy Policy*, vol. 38, no. 11, pp. 6439–6452.

Zhou, YX 2016, *"Qifeng xiandian" zhi yinxing nengyuan guisun da gudong chengnuo shiyan yigai buchang fang'an ("Restricting wind power" led to profit deficit of Yinxing Energy, big shareholders admit breaking promise and drafting compensation plan)*, viewed 02 June 2016, <http://finance.sina.com.cn/roll/2016-04-14/doc-ifxriqqx2366914.shtml>.

6 Updates during the 13th Five-Year Plan and implications for unaddressed issues

Introduction

On 5 January 2017, the State Council of China announced the *Comprehensive Work Plan for Energy Conservation and Emission Reduction Scheme (ECERS) during the 13th Five-Year Plan*. It identified the new status of China's economic development characterised by the slow-down of energy consumption growth as well as the slow-down of energy-intensive, resource-intensive, and high-emission industries. The new status provided a fortuitous opportunity for the optimisation of China's industrial structure. But China's overall energy demand continues growing. The resource and environmental problems are still one of the biggest restrictions on China's economic development. Therefore, China's energy conservation and emission reduction still faces tough challenges (State Council of PRC 2016, n.p.).

The new *Work Plan* confirmed the importance of roles played by the government, the industry, the market, and the public in energy conservation and emission reduction. It also stressed the urgency of improving the incentive- and punishment-based system, ascertaining the responsibility of the provincial and local governments, and strictly implementing the environmental laws and regulations. Moreover, the *Work Plan* proposed to further promotion of market mechanism and incorporating economic, legal, technical, and necessary administrative tools in energy conservation and emission reduction (State Council of PRC 2016, n.p.). These directions undoubtedly echoed the policy weaknesses identified in the analyses carried out in previous chapters during the 11th and 12th Five-Year Plans. However, it did not include specific instructions on how to address them. So, this chapter investigates whether and to what extent the identified policy weaknesses of the 11th and 12th Five-Year Plans are addressed during the 13th Five-Year Plan. For ones that are not dealt with, this chapter will explore possible implications.

New updates and recommendations for China's climate change actors

Chapter 3 analysed the "who" factor in Chinese climate-related policies. The government, the think tanks, the enterprises, and the civil society were all involved in China's climate affairs to varying degrees. As an authoritarian state, the Central Government of PRC undoubtedly played the leading role in the making of climate-related policies. But there are several issues that needed to be addressed in the government actor in order to improve the coordination and to develop sensible and effective national climate policies, namely the overlapping authorities among relevant departments within National Development and Reform Commission (NDRC), conflicts between NDRC and Ministry of Environmental Protection (MEP; now Ministry of Ecology and Environment, MEE), and competition of interests between MF and NDRC. The expert community played an increasingly important role in China's climate-related policy formation. Expertise had been absorbed from broader sources including non-official and overseas experts. However, expertise in climate change mitigation still needed to be strengthened at the provincial level. Energy enterprises were an important actor who were not just the implementers of relevant climate-related energy policies, powerful state-owned enterprises (SOEs) also influenced the formation of policies through bargaining with the government. Newly emerged private renewable energy companies also played an increasingly important role in climate change mitigation and the research and development of climate-friendly technologies. Therefore, it was important to provide incentives and support to the energy industry so that they are confident for more ambitious emission mitigation. The media pushed forward the public debate on air pollution and climate change, which was a pioneer force for awakening the public's environmental awareness. However, it faced much political pressure and restriction in reality. The media watch dog would only effectively guard the environmental integrity if this situation turns around. The involvement of the environmental non-governmental organisations (ENGOs) and the general public was increasing with the broadening usage of internet and social media, but was still limited in China's climate affairs. These issues all need to be addressed in order to achieve a better outcome in China's energy conservation and emission reduction. The following sections will follow up any new updates and draw out some useful implications for these issues.

New updates for China's environmental ministry

Among all climate-related actors, the government actors have undergone the most significant reform in the 13th Five-Year Plan. Chapter 3 of this book identified a range of weaknesses within China's climate-related government actors during the 11th and 12th Five-Year Plans. For instance, there were overlapping tasks among three relevant departments within NDRC: Department of Climate Change (DCC), National Energy Administration (NEA), and Department of Resource Conservation and Environmental Protection (DRCEP). There were also serious conflicts between MEP and NDRC and competitions between MF and NDRC. These issues were addressed to various extents during the 13th Five-Year Plan. In March 2018, NDRC announced a transfer of a range of its responsibilities to its brother ministries. Among these, NDRC's planning and organising functions on major resources were transferred to the newly established Ministry of Natural Resources (MNRs). The MEP was renamed MEE. And NDRC's management function of climate change was moved to MEE. On 8 October 2018, the Department of Climate Change (DCC) was formally placed under MEE (Huo 2018, n.p.). By then, MEE included the most significant greenhouse gas, CO_2, in its management portfolio. Climate change was no longer managed by the economic development driven ministry (Huo 2018, n.p.).

The reform has its significance in terms of balancing the power between China's environmental and economic ministries. This is because many of DCC's functions have the capacity to produce positive environmental outcomes. DCC covers a wide range of responsibilities of climate change adaptation and greenhouse gases emission reduction. It is responsible for (1) conducting analysis of the impact of climate change on social-economic development; (2) organising and implementing national climate change mitigation strategy; (3) composing major targets, policies, regulations on greenhouses emission reduction, green low-carbon economy and climate change adaptation, and coordinating relevant departments, industries, and local authorities for effective implementation; (4) leading China's compliance of UN climate change framework and organising and attending international negotiations and conferences together with other relevant departments; (5) promoting bilateral and multilateral cooperation and communication, conducting climate change mitigation capacity building, research, and promotion; (6) organising and implementing CDM; (7) managing national ETS; (8) fulfilling detailed tasks of the National Leading Group for Climate Change; and (9) fulfilling national tasks for international agreement on ozone protection (MEE 2018, n.p.).

Among all of DCC's responsibilities, Articles (2) and (3) clearly state its leading role in climate change policy development. Due to the fact that CO_2 is the most significant greenhouse gas that impacts all aspects of the national economy, DCC must be actively engaged with the economic sectors, the industries, the market, and the general consumers. This means that MEE starts gaining the rights for involvement in the strategic planning of some of the nation's carbon-intensive industries and emission reduction programs. This is one step forward in term of strengthening China's environmental ministry. MEE truly starts to hold some solid power compared with MEP which was described by its own staff as merely "a meter at the end of the pollutants discharge drain" (Interviewee 3).

However, whether the power gain of MEE is enough to pull back China's development-driven economic pattern remains to be seen. Back in October 2013, Interviewee 1 from NDRC firmly believed that "MEP has no capability of managing climate change, even if NDRC passes on the authority to them [MEP]". This is because CO_2 is unlike other types of greenhouse gases. It involves the every aspect of the national economy. Five years later, this statement remains a concern for the new MEE. For MEE to truly bring positive national emission reduction outcome, it not only needs to be precise in emission reduction policies, but also radiate enough influence on major national development plans so that to shape emission reduction outcome right from the beginning. Currently, industrial development planning and major national programs are under the supervision of NDRC. For instance, the power industry is managed by NEA under NDRC. This is to say, if NDRC plans to set up a range of energy and emission intensive programs, MEE, although it now has the power to question, still needs to work out how to proceed with it. In this sense, there is still much to be down for China's environmental ministry to balance the power of the economic ministry. Fundamentally, for MEE to manage climate change well, it requires MEE to have power to influence or even constrain NDRC's decision-making on projects that poses detrimental effect for China's long-term climate change mitigation and overall environment.

Nonetheless, the recent government reform is positive for the overall integrity of China's emission reduction scheme. MEE has started to work as an auditing body for many industries and companies in climate change mitigation programs. The expertise of MEE in monitoring and auditing non-CO_2 emissions is mature and advanced. Therefore, when the production material and technology of an industry or a particular company are certain, the auditing results on non-CO_2 emissions will reflect its CO_2 emissions. China's carbon emission reduction programs are no longer managed in a single dimensional manner led by NDRC and checked by NDRC. The carbon emission reduction results will therefore be more objective and convincing, which in turn will lead to improved transparency, accountability, and increased chances of policy success (Gao & Zhu 2010, p. 5).

To enable MEE for the above function, the 13th Five-Year Plan also proposed that provincial MEP (now MEE) carry out direct supervision for environmental auditing and law enforce of its local branches. Although the provincial and municipal offices of China's central ministries are generally managed by provincial and local governments, it is reasonable for MEP (now MEE) to become an exception. As mentioned in Chapter 3, provincial and local authorities' protection of polluting industries and their control of local Environmental Protection Bureaus (EPBs) were great obstacles to environmental protection. Therefore, provincial EPBs providing direct supervision to their local EPBs was a positive innovation due to China's dramatically worsening environmental status and rapidly increasing carbon and non-carbon emissions. The proposal planned to complete the pilot program in 2016, and complete the national reform before the end of 2018. MEP also provided direct funding to EPBs and was responsible for recruiting EPB officers in this initiative (Wang 2016, n.p.). It would be a bold reform for MEE to direct supervise all national branches which requires huge resource input into China's environmental sector. However, such a reform will prove worthy in the long run as China's environmental and climate change issues deserve top priority.

WG approach for overlapping tasks among DCC, NEA, and DRCEP

DCC was already relocated under MEE, however, the other two climate-related departments, NEA and DRCEP, are still under the management of NDRC. The relocation does not automatically solve the overlapping tasks among the three departments responsible for energy and climate policies. The work of NEA, DCC, and DRCEP all have their specialty but are also interconnected. For instance, NEA focuses on national energy planning and energy intensity reduction, DRCEP emphasises on energy conservation, and DCC works on greenhouse gases mitigation and uses carbon intensity as a main indicator. The three departments mainly conducted their work individually. They produced three similar sets of documentations with slightly different criteria, which increased the workload at both central and provincial levels. The lack of interaction among the three departments dramatically reduced efficiency.

In order to improve efficiency, this book suggests a whole-of-government (WG) approach. WG, earlier known as joined-up government, was first proposed by the Tony Blair government in 1997. It aimed at tackling complicated problems and policies across departmental boundaries (Richards & Smith 2006, p. 187). According to the report *Connecting Government: Whole of Government Responses to Australia's Priority Challenges* conducted by the Australian Management Advisory Committee, "whole of government denotes public service agencies working across portfolio boundaries to achieve a shared goal and an integrated government response to particular issues" (Australian Public Service Commission 2004, n.p.). The WG approach can be delivered in a formal or informal manner across traditional departmental and hierarchy boundaries, but there has to be obvious benefits such as efficiency improvement and resource saving. The WG approach is not usually used for routine and straightforward issues. Instead, it is more suitable for complex and persisting policy issues, or the "wicked problems", which often challenge the traditional bureaucratic boundaries and routines (Australian Public Service Commission 2004, n.p.).

The WG approach is a suitable approach for the Chinese government in the development of climate-related energy policies. This is, first, because WG can optimise both horizontal coordination among central ministries and vertical coordination between central ministries and their provincial and municipal branches without shovelling around the existing bureaucrats (Christensen & Lægreid 2007, p. 1058). In a system where overlapping exists and multi-dimensional coordination is required, WG can provide the solution. Second, the Chinese government is capable of pursuing a WG approach. The WG approach requires a strong central government for effective coordination and supervision (Richards & Smith 2006, p. 187). For China, the strong central control enables the implementation of WG. In fact, there are already elements of WG approach on climate-related issues within the Chinese government such as the establishment of National Leading Group on Climate Change which chaired by the premier and included most ministers. However, the National Leading Group's work is more symbolic than practical. WG in climate change and the energy domain needs to be applied to detailed policies and programs, and specific operating procedures of the approach should be defined and made clear to relevant ministers and departments.

It is recommended to establish a framework that enhances departmental coordination. An evidence base for this approach can be drawn from Australia. The Council of Australian Governments (COAG) institutional framework is a successful example which provides good initiatives towards the WG approach. The COAG hosts a forum twice a year for intergovernmental collaboration and decision-making on key national priorities. Mid-level public servants also meet and collaborate in between these across all departments (COAG 2017, n.p.). The National Water Initiative (NWI), for example, is one WG approach towards the better management of Australia's water resources under COAG. It included more than 40 Commonwealth-state ministerial councils and worked effectively towards improving water usage efficiency, directing investment, and balancing between the needs of the communities and the environment (Department of Agriculture and Water Resources, Australian Government 2017, n.p.).

For China, it is recommended to establish a similar institutional framework towards the WG approach. The National Leading Group on Climate Change already fulfils similar function to the high-level forum of COAG. However, there is need for setting up a network of smaller forums with each focusing on its special tasks. By doing so, the forums can stay focused on individual policy areas. The climate change forum could be one component of the national WG framework. It can provide the opportunity for the climate-related central departments and provincial governments to work together and conduct collaborative work plans. Specifically, the three departments mentioned above should produce a more comprehensive report on China's overall energy, energy conservation, and climate change mitigation plan through such a framework. For instance, NEA and DCC can make co-decision on the national energy plan and emission reduction target for the energy sector. Based on these targets, DRCEP can join in to work out detailed energy conservation target, and can work especially closely with DCC for emission reduction strategies. The collaboration can be carried out through a series of formal and informal meetings. Formal meetings are planned with set schedules where other stakeholders such as energy and climate change experts, industry leaders, and personnel from other ministries are also invited. Informal meetings such as video meetings and email communication on specific policy details can happen between responsible officers of these three departments as often as needed.

Solutions for competitions between MF and NDRC

The competitive relationship between MF and NDRC on climate change during the 11th and 12th Five-Year Plans was mainly around the argument of whether to implement carbon tax or carbon trading in China. In this battle, NDRC definitely won. In December 2017, NDRC announced the *Construction Plan for the National Carbon Emission Trading Market (the Electric Power Industry)*, which indicated the official launch of China's national carbon market. But there is still discussion on the possibility of carbon tax in the 13th Five-Year Plan. According the Jiang Kejuan, the researcher from Energy Research Centre, there were originally three pages of discussions on the methods of carbon tax operation in the *Environmental Protection Tax Law* promulgated on the 1 January 2018. However, this content was later deleted due to the lack of unanimous decision. The main argument laid on whether carbon tax was meaningful when national ETS was already implemented. Jiang Kejuan personally believed that carbon tax should still have its role. This is because carbon tax is more practical in terms of operation. Carbon trading relies on data and auditing; without these elements, carbon trading is very difficult to carry out. The coal-fire power industry, as the first industry included in the national ETS, is of good shape in terms of data's comprehensiveness and accuracy, thus making auditing relatively less complicated. However, not all industries can provide this. The more industries entering the ETS, the tougher the implementation. In contrast, database is not essential for carbon tax, as a result, there is easier operation and cost reduction on auditing (Zhu 2018, n.p.). Associate Professor Wang Ke from the Energy and the Environmental Policy Centre, Beijing Institute of Technology stands for the middle way that carbon tax and carbon market are not contradictory. Instead, they should complement each other. Currently, only enterprises with an annual CO_2 emission greater than 26,000 tce or total energy consumption greater than 10,000 tce are included in the national carbon market. For the rest of the enterprises with smaller emissions, carbon tax can be charged as supplementary to the national ETS. Jiang Zhaoli, the vice director of DCC also gave out signal on the 2016 China Carbon Market Outlook Conference, that after 2020, carbon tax might be considered for enterprises with energy consumption lower than 5000 tce or enterprises that are not included in the national ETS (Zhu 2018, n.p.).

According to signals from various relevant government departments, carbon tax is to be a post-2020 consideration. Therefore, it is beyond the scope of the analysis of this book. As carbon trading is already chosen as China's preferred emission reduction strategy, China's climate-related government departments should work together to support and improve the current carbon market. The competition for the authority over climate change should be over, especially when now DCC is under the management of MEE. MF should accept this status and cooperate with MEE in optimising China's national carbon trading system. In order to provide enough incentive for MF to get involved in the ETS, MEE can involve MF in financing and finance auditing tasks. This is because MF has expertise in these tasks, which appear the most challenging of China's current carbon market. Such integration enhances the WG approach. It can be expected that MEE and MF cooperate more in future to deliver joint programs beneficial to China's climate change mitigation and overall environmental outlook, such as the conversion of different types of pollution surcharge and fees into taxes.

Overall, China's climate-related government sector has undergone multiple adjustments during the 13th Five-Year Plan. The most significant one is the relocation of DCC from NDRC to the new MEE. This strengthens China's environmental protection and enables MEE to gain some solid power. The initiation of direct supervision of local EPBs broadens MEE's authority on environmental governance. The provincial and local EPBs are expected to be better funded and less restricted by the provincial and municipal governors when conducting pollution-related investigation. There is also possibility for MEE and MF to co-develop carbon tax after 2020 as a supplementary policy to the national ETS. However, frequent and timely meetings between relevant departments should be held when necessary to conduct in depth discussion over policy development and program supervision in order to avoid overlapping and inefficiency.

Strengthening climate change expertise at the provincial level

The lack of climate change expertise and research capacity at the provincial level has been an ongoing problem. Beijing is obviously the best in terms of climate change research expertise and personnel. For instance, the Beijing Climate Change Response Research and Education Center (BCCRC), jointly established by Beijing City NDRC and Beijing City Education Committee in 2012, fulfilled a wide range of research and expertise building responsibilities on climate change mitigation (BCCRC 2012, n.p.). However, organisations similar to BCCRC are not common at provincial and local levels across the country. There has not been significant development in similar organisations during the 13th Five-Year Plan. This book believes the quickest fix to this problem is to provide professional development for existing public servants in areas of climate-related policies and implementations. According to Lin and Gil (2016, p. 148), customised training and examinations are effective ways for Chinese public servants to enrich their knowledge and capacity in the environmental protection sector. This book also encourages the provincial government to liaise with scholars and intellectuals from local universities with relevant expertise. Expert knowledge and advice have proved beneficial for the government's policy development and implementation. For instance, the municipal government of Cape Town, South Africa, liaised with researchers from the Energy Research Centre of Cape Town and conducted successful studies in reducing peak energy demand. The New York City government also adopted similar procedures by including talented professionals in the New York Panel on Climate Change and raised a range of initiatives in reducing carbon emission (Anguelovski & Carmin 2011, p. 171).

Furthermore, courses and topics on climate change can be provided at universities so that there will be a sustainable supply of graduates to join the climate-related sector. For instance, climate change related topics can be added to the university courses of environment, policy, and science majors. Famous scholars with climate change expertise from home and abroad can be invited to provincial universities to provide guest lectures to students in this discipline, since China has a long tradition of employing foreign experts in areas such as science and education. Relevant provincial government departments and research institutions should provide graduate recruitment opportunities to university students in climate-related disciplines in order to absorb talent and expertise in climate change policy research. Training and certificate in carbon emission monitoring, trading, and administration can also be provided for people who wish for a career change or who already work in the climate change field and wish to enrich their knowledge. By providing customised professional development to public servants, absorbing scholars and talented stakeholders, enriching university curriculum and society recognised training, climate change expertise and research capacity can be quickly built up at the provincial and local level. This build-up of expertise will not only better serve climate change mitigation programs at these levels, but will also better correspond to the central government's climate policy and providing more rational feedback to assist central government's decision-making.

Ways for energy corporations to carry out ambitious emission reduction

China's energy corporates are important players for its national emission reduction scheme. Major large energy SOEs hold the key for China's energy intensity reduction as they took charge the biggest proportion of China's coal-fired energy production. Private renewable energy corporates have become the pioneers in research and development in renewable energies and a new force in optimising China's energy structure. Large energy SOEs, although willing to accept emission reduction procedures such as shutting down inefficient small coal-fired power plants, lack of incentives to pursue more ambitious emission mitigation such as bigger investment in renewable energy. Medium- and small-scale private renewable energy enterprises, although ambitious, face many obstacles such as lack of finance and late and unpaid subsidies from the government. This book recommends establishing alternative incentives for large energy SOEs and providing more support for private renewable energy investors, as will be discussed below.

The analyses of the 11th and 12th Five-Year Plans showed that large energy SOEs only passively included renewable energies in their business portfolio under the government's administrative orders. Administrative orders alone were not sufficient for large energy SOEs to adopt more ambitious carbon emission reduction plans. It was true that SOEs were state-owned; however, they were still involved in market competition and were profit driven. For instance, almost all large energy SOEs preferred large-scale coal and gas projects because they were competitive and most profitable under the market conditions of the analysed period, when carbon emissions were not priced. It is widely agreed on that to make large energy SOEs truly willingly embrace renewable energy and make renewable energy an important and growing part of their business, it is essential to reduce the profit margin of fossil-based energy to an extent that renewable energy is more economically attractive.

At the end of 2017, China's national ETS was officially launched. This is China's most significant move of the 13th Five-Year Plan in climate change mitigation. The electric power sector, especially the coal-fired power, was first to be listed into the national trading portfolio. To put a price on carbon emissions will make renewable energy more attractive and competitive, theoretically. However, according to the analysis carried out in Chapter 5, simply making renewable energy more price-competitive was not enough for large energy SOEs to truly embrace renewable energy. This was because the more competitive renewable energy price would not automatically remove the obstacles that renewable energy currently faced such as the politics played between provincial governments and the coal mining industry. Therefore, a second tier of market guarantee from certain energy consuming sectors is needed on top of carbon trading.

The Chinese government should actively promote new market sectors that consume only renewable energies. This can be achieved through legislations. For instance, it could be regulated that China's electric car recharging facilities must only purchase and sell electricity generated from renewable sources; real estate development proposals with electric car recharge facilities have priority in gaining approvals; and owners of electric cars do not need to be put on a waiting list for registration in big cities like Beijing and can drive any day rather than alternate days like petrol car owners. In the northern provinces, winter central heating must consume locally produced wind-energy before using coal.

These market end regulations will provide a second tier of protection for renewable energy against the dirty politics played between some provincial governments and coal mines. These will also provide more convincing incentives for large energy SOEs to actively develop renewable energy and set more ambitious emission reduction targets.

The market end guarantees will not only provide incentive for large energy SOEs, but also encourage medium- to small-scale renewable energy investors equally. However, for medium- to small-scale renewable energy to be competitive, extra supports are needed. First of all, the sources for renewable energy subsidy should be diversified rather than solely relying on the renewable energy surcharge. As explained in Chapter 5, the renewable energy surcharge alone is not sufficient to support the whole renewable energy industry. The late payments of subsidies have put many medium- and small-scale renewable energy companies in financial difficulties. To diversify the sources of the renewable energy subsidy, a special financial fund can be established for medium- to small-scale private investors in renewable energy. This fund could be generated from the government's auctioning of carbon emission credits. Under the national ETS, this fund will be a useful source to guarantee sufficient subsidy and other financial support. Medium- and small-scale renewable energy investors will be relieved from financial pressures caused by unpaid or late subsidy.

Tightened control for Chinese media under Xi's leadership

The media has been an important actor in China's climate-related affairs. They work on reporting pollution problems and generating discussions to awaken people's environmental awareness. However, in the 13th Five-Year Plan, the political restrictions over the media were not removed or relieved. In fact, there were signs of tightened media censorship under Xi's leadership (Ng 2013). In February 2016, President Xi Jinping announced a new media policy for the party and state news, which stated that "all the work by the party's media must reflect the party's will, safeguard the party's authority and safeguard the party's unity" (Xu & Albert 2017, n.p.). Other media channels such as Chinese internet companies were not only required to follow the *2010 White Paper* which detailed the laws and regulations on China's media industry, but now are also required to sign the *Public Pledge on Self-Regulation and Professional Ethics for China's Internet Industry* (Xu & Albert 2017, n.p.). The tightening up of media censorship of both the state media and the internet showed the media sector faced even tougher challenges when reporting serious, large-scale environmental issues, especially those involving a large population and having broad influence. These measures weaken the media's role as an effective guardian for environmental integrity. Therefore, it is a shame to say that the media actor did not progress as wished in the current Five Year Plan. No similar inspiring documentary like "Under the Dome" has appeared in the public sphere ever since.

Strengthening China's civil society

The involvement of Chinese ENGOs and the general public in China's climate change policy is limited. However, as discussed in Chapter 3, due to the interconnection between climate change and China's worsening atmospheric pollution, growing attention from the general public has been observed. The government should realise that the atmospheric pollution problem is already a nation-wide problem which has caused much complaint from citizens. Many of the rich have chosen to migrate to other countries where environmental conditions are optimal. They bring along large amounts of wealth to the destination country which causes the loss of wealth to China (Cendrowski 2014, n.p.). The poor have no financial capacities to do so and therefore have to bear the high levels of pollution. As pollution levels increase and related sicknesses intensify, the poor who suffer from the diseases have no money to get proper medical treatment, atmospheric pollution quickly transforms from an environmental problem to a serious social problem which may affect social stability – something the Chinese government is most concerned about. The analysis of Chapter 3 confirmed that the political environment for Chinese civil society would be unlikely to change in the foreseeable future and China's climate change and atmospheric pollution problems were almost certain to be solved through a top-down approach. Under such circumstance, it requires the Chinese government to take these problems seriously and view the civil society as a positive and effective actor to fight against climate change and atmospheric pollution. This is because the media, the ENGOs, and the general public can quickly identify local cases of pollution, which provide valuable clues for local EPB's investigations.

In order for Chinese ENGOs to carry out effective activities against illegal polluters within the set political limits, such as the pursuit of civil environmental lawsuits mentioned in Chapter 3, the government should establish a transparent and accessible environmental legal system by updating relevant environmental laws and simplifying environmental lawsuit procedures. In 2015, MEP set up the 12369 Environmental Protection Hotline where citizens could report polluters anonymously. EPBs would conduct initial investigations to confirm the truthfulness of the cases and deal with them accordingly. The public greatly welcomed and valued the hotline, however, the outcomes of different cases in different regions turned out quite differently. Some EPBs dealt with the public report in a timely and professional manner while others did not due to a range of practical difficulties such as lack of staff and funding (Zhihu 2015, n.p.). In order to overcome the practical difficulties some EPBs encountered, this book suggests EPBs share some of the cases they receive from the hotline with a range of local ENGOs. Such cases should be of less complexity and within the capacity of local ENGOs. Specifically, the information of the cases can be published on the EPB website in the form of a reward list with clearly stated cash prize amounts. Usually the cash prizes would be smaller than the expenditure of EPBs conducting the investigation by themselves. Interested ENGOs can register to take over the cases they wish to investigate, and then conduct thorough investigations of the pollution according to the EPB's standard procedures. EPBs will reward the ENGOs with the promised funds once the cases are completed. This form of subcontracting will not only reduce EPBs' workloads, but will also encourage ENGOs' enthusiasm to conduct more timely investigations against polluters.

Some may argue that such an initiative is very unlikely given the closed government in China. However, two conditions have been stipulated for this recommendation. The first is that it operates within current political limits. The second is that the cases to be shared are within ENGOs' capacity and of lesser complexity. These two conditions can provide cushioning to the local EPB and ENGO relations. Although the Chinese government is often viewed as a closed government, it is generally willing to experiment with new initiatives and is generally quick to embrace them if there are obvious benefits. The pilot carbon trading program and the newly landed national ETS are such example. Therefore, the adaptive characteristic of the Chinese government should not be overlooked. Furthermore, successful cooperation between local EPBs and ENGOs have been carried out elsewhere in the world, which further demonstrates the feasibility of this approach. For instance, the government of Melbourne paired traditional command and control governance with an NGO, the Moreland Energy Foundation, in conducting climate change mitigation which aimed at reducing waste and improving household energy efficiency. In Quito, the capital city of Ecuador, the environmental office provided funding directly to local ENGOs to carry out climate change adaptation training to local farmers in water management and urban agricultural practices (Anguelovski & Carmin 2011, pp. 171–172). Both cases achieved positive outcomes. Therefore, it is believed the potential cooperation between Chinese EPBs and qualified local ENGOs will have the capacity to bring better results to local environmental governance.

It is also advisable that ENGOs set up public blogs and use social networking software such as WeChat and Weibo to connect to the general public. In this way, the public has a channel to report local polluters to local ENGOs and the relevant ENGOs can conduct investigations and sue polluters through the legal system. Because the cases are handed to ENGOs with a mission of fighting environmental offences, they are therefore more likely to be dealt with quickly.

Consumers in China should also have the choice of purchasing green energy. In China, renewable energy costs are shared by all end users, which is to say all end users pay the same price for energy within a certain region. Consumers do not have the choice of purchasing only green energy or a certain percentage of green energy. Although it is likely that the majority of consumers would choose to purchase the cheapest energy option, there will be some environmentally conscious consumers who are willing to pay for green energy. Under the national ETS green energy will be more price competitive. As the education level of China's younger generation on the rise, the message of eco-civilisation is widely spread. Also, as people's living standard improves, there should be an increasing number of consumers in this group (Wang & Yang 2014, n.p; Zhou 2014, n.p.). This situation provides scope for the diversification of the price of renewable energy.

Specifically, renewable energy generating companies can promote their products through multiple channels such as TV or internet advertising, ENGOs' environmental education sessions, and recommendations of energy providers. Energy providers can introduce this new green energy product to their customers and sign contracts with those who are interested. The profit gained from the new product will be distributed between the green energy producer and the energy provider in the proportion stated in their contract. This option of using green energy will provide people with a sense of fulfilment, and make them feel good as they are doing a good thing for the environment and future generations (Hartmann & Apaolaza-Ibáñez 2012, p. 1254).

Summary of actor updates in China's climate-related energy policies

There has been fractured growth among China's climate-related actors during the 13th Five-Year Plan. The government sector has obviously undergone a positive move towards a more rational organisation of its climate change governance, namely moving DCC under the supervision of MEE. The media, on the contrary, experienced tighter control under Xi's leadership, especially after Xi emphasised the close relationship between the nation's media outlets and political stability. It is a shame because a less free media means a lot more reliance on the government's self-regulation. Whether MEE could promptly detect, be willing to address, and let the public know of pressing environmental problems remains to be seen. The industries and the expert community basically retained their status of the 11th and 12th Plans period. Based on the updates and analyses, it is recommended that the climate-related departments should adopt a WG approach to carry out better communication and coordination. The expert community should spread their climate change expertise into the provincial and municipal level through tertiary education and training program. Renewable energies can be protected by the promotion of certain green-energy-only sectors such as electric car charging facilities and winter heating in China's northern provinces. Medium and small investors are recommended for extra financial support generated from the fund of auctioning of carbon credits. China's ENGOs can be strengthened by improving environmental laws and an easier legal process for environmental lawsuits. An innovation between EPBs and ENGOs can be carried out by using subcontracting system to deal with polluters. ENGOs should also create more channels for the general public to participate in local environmental discussions and to report suspicious polluters. The general public should also be provided with more diverse options for their energy sources.

Updates and implications for issues raised in Chapter 4

Chapter 4 in this book analysed China's participation in international carbon offsetting program CDM and its impacts on China's sustainable development and national carbon market during the 11th and 12th Five-Year Plans. It confirmed that CDM had brought multiple benefits to China's sustainable development such as raising climate change awareness, promoting renewable energy development, and preparing for China's domestic carbon market. There were also a few issues that need to be addressed. First, how should the unsold CERs from unilateral CDM (uCDM) projects be dealt with? Second, how can China guarantee the additionality of Chinese Certified Emission Reductions (CCERs)? Third, how should the involvement of China's domestic financial institutions in the carbon market be encouraged? Finally, what low-cost climate change abatement options are left after CDM in China? This section will focus on the investigation of these questions and discuss their implications.

uCDM projects and the urgent need for national CCER regulation

As mentioned in Chapter 4, uCDM projects had a number of advantages over bilateral and multilateral CDM projects, such as lower transaction and administrative costs, the ability to quickly develop a range of small-scale projects, and the control of the hosting country over the sales of CER. uCDM projects also faced some challenges. One of the challenges some uCDM faced was the inability to find international CER buyers or selling CERs at a greatly discounted price due to the downturn of global carbon market. In order to solve this problem, Interviewee 1 from NDRC indicated that there is the possibility of absorbing CERs from such uCDM projects into China's domestic carbon market, and renaming this type of CERs CCERs (September 2013). The domestic pilot carbon trading program accepted four types of CCER projects. Two types of projects had already been mentioned in Chapter 4. They were CDM projects approved by NDRC but failed to register with CDM Executive Board (EB) and CDM projects successfully registered with CDM EB but failed to find CER buyers. The third type was voluntary emission reduction projects that followed and passed NDRC approved methodologies. The last type of CCER projects were those that gained NDRC approval for CDM and produced CERs before registering with EB (Shen 2015, n.p). CCERs from the above types of uCDM projects had been generated and used as a supplement to trading permits in China's pilot carbon trading program and could be purchased by relevant industries and companies to offset their carbon emission (Lo & Cong 2017, p. 1395).

It is expected this offsetting mechanism can significantly lower the abatement cost for companies which would otherwise spend more to achieve efficiency improvement and emission reduction from internal technological updates. By now, the CCER projects as a continuation of international CDM have already been identified as an important component of China's national ETS. This is to say, the project-based mechanism will continue to have its role in China's national carbon market (Gonzalez 2014, n.p.).

2015 was the first year for CCERs to enter China's carbon trading pilots. However, the implementation of CCERs was not at all smooth. During the 2016 compliance year, there were only 8 million tCO_2e used for carbon offsetting due to delayed policies, which was far below the preset proportions. In March 2017, NDRC announced it would temporarily stop the assessment and approval for CCER projects. The announcement directly led to a CCER price drop. For instance, the CCER price in Shanghai dropped to below 15 yuan, while in Beijing the CCER price dropped to below 10 yuan. The price drop was due to the uncertainty of CCER usage in China's upcoming national market (Gong 2017, n.p.). However, NDRC's announcement was necessary as there were indeed a range of issues which needed to be addressed before the implementation of the national ETS.

An urgent issue associated with CCER was the lack of unified national regulation. Up until 8 April 2018, CCER Info-Platform announced 2871 registered CCER projects, among which 287 projects had gained approval and 254 projects had already produced 53 million CCERs (Tanjiaoyi 2018, n.p.). But which of these CCERs could be used in the future national carbon market was not clearly stated. During the 2013–2015 experimental period, different carbon trading pilots had different restrictions on the CCER type, time frame, and origin. For instance, the Beijing market required that, to be qualified, CCERs had to be generated after the 1 January 2013 and could not be from industrial gas projects such as HFCs, PFCs, N_2O, and SF_6 (21st Century Economic Report 2016, n.p.). All pilots except Shanghai excluded CCERs generated from hydro-power or at least excluded large- to medium-scale hydro-power due to its significant impact to the environment. Also, Chongqing and Shanghai were the only two pilot areas that did not have restrictions on the origins of CCERs. All the rest of the pilots required majority or partial CCERs to be locally generated (Zhongchuang Tantou 2015, n.p.).

These different CCER regulations were catered for different cities and regions, but, as of for the national carbon market, a new national CCER regulation is urgently needed. On the occasion of China Carbon Trading Open Day, held on 15 April 2016, the Deputy Director of Domestic Policy and Compliance Division, DCC, Wang Shu, mentioned that the regulation and assessment for CCER projects of the national carbon market would be stricter and tougher. In future, the proportion of CCER projects would be restricted, but the time frame for qualified CCER projects was yet to be decided. Whether the national carbon market should accept CCERs produced after 2013, 2014, or 2015 is an issue which required consultation with the experts. Wang also indicated that earlier projects were likely to be excluded from the national carbon market and recommended investors to focus on new projects with a higher contribution to sustainable development. Industrial gas projects were likely to be excluded; large hydro-power projects might be reconsidered for the national CCER category (21st Century Economic Report 2016, n.p.). However, by the end of 2018, the assessment and approval of CCER projects had not been resumed.

This book suggests that CCER projects as an important offsetting component for China's national ETS should be given clear guidance, and that the regulations should be made in a timely manner in order to properly facilitate investment and truly embrace environmental integrity. This is because not only do potential CCER investors need clear guidelines, but corporates, which participate in the national ETS, also need time to investigate their targeting CCER projects. It is strongly recommended that the development of national CCER regulation should strictly follow the principles of sustainability and additionality, provide informative and supportive financial service to CCER projects, and be innovative in abatement options.

Principle of sustainable development and additionality

A qualified CCER project must contribute to sustainable development and pass the additionality test. As mentioned previously, the Chinese government had the intention not to include industrial gas projects in its CCER portfolio. This was a positive decision because industrial gas CDM projects contributed little to sustainability. However, the reconsideration of large hydro-power is not recommendable. This is because large-scale hydro-power developments often pose very negative environmental and social impacts to the local ecosystem and communities. China's Three Gorges Dam, for example, raised numerous problems during its development. The diversity and abundance of both flora and fauna in the region were seriously impacted due to habitat change (New & Xie 2008 p. 3157; Xie 2003, p. 1149). The built up concentration of water pollutants threatened the health of the East China Sea ecosystem (Mueller et al. 2008, p. 245). The negative social impacts were no less than the ecological trauma. Over 1.3 million people were relocated causing loss of farmland, unemployment, and deteriorating public health (Jackson & Sleigh 2000, p. 223). The even more worrying factor was the potential risk of large earthquakes associated with the construction of the dam and the frequent fluctuation of water levels. Landslides and regional dam quakes already happened in the regions causing loss of properties and lives. Large-scale dam quake cannot be excluded from the possibilities (Hvistendahl 2008, n.p.).

With negative social and environmental impacts like this, it is unethical to justify the gains of large-scale hydro-power projects solely from the carbon emission reduction perspective; and obviously its contribution to sustainable development is doubtful.

Large-scale hydro-power projects are also questionable in terms of their additionality. Additionality is the most fundamental criteria that an offsetting project needs to meet. If China's carbon market lets in CCER projects failing to justify their additionality, it harms the integrity of China's entire ETS. The Three Gorges Dam was planned and constructed long before climate change mitigation was on the Chinese government's agenda and had little relevance towards CO_2 emission reduction. There were other large-scale hydro-power projects which followed a similar pattern and were primarily economically driven, which would have happened anyway even if the CCER policy was not in place. Therefore, it was difficult to justify their additionality. It is irrational for the national ETS to grant CCERs to large hydro-power proposals especially when considering their potential negative social and environmental impacts. It is recommended to exclude large hydro-power projects from the CCER project category in order to safeguard environmental and climate integrity. Instead, China should explore new CCER methodologies for smaller projects with greater contribution to sustainable development and clear justification to additionality.

Exploring more low-cost climate change abatement options

China was active in the exploration of new low-cost carbon abatement options. The Chinese government encouraged domestic consultancies and research institutions to develop new CCER methodologies since the start of China's pilot carbon trading program. In June 2012, NDRC announced the rules for CCER projects which closely resembled the procedures of CDM (DCC, NDRC 2012, n.p.). Since then, NDRC adopted all 173 CDM methodologies and developed 30 new methodologies (China Certified Emission Reduction Exchange Info-Platform 2017, n.p.). This outcome echoed the analysis in Chapter 4 that China still has the potential to unearth new low-cost abatement options after hosting the majority of international CDM projects.

Many of the newly added CCER methodologies targeted smaller projects which focused more on sustainability (see Table 6.1 for full list). For instance, a similar project methodology to the small-scale eco-recovery projects proposed in Chapter 4 had been jointly developed by the Nature Conservancy and Beijing Environment Exchange and approved by NDRC on 15 January 2015 (Methodology ID: CM-099-V01). A range of methodologies for afforestation/reforestation projects were also developed (Methodology ID: AR-CM-001-V01; AR-CM-002-V01; AR-CM-003-V01; AR-CM-005-V01; Table 6.1).

Table 6.1 New CCER methodologies approved by NDRC

Methodology ID	*Name*	*Time of approval*
AR-CM-001-V01	Carbon sink afforestation	4 November 2013
AR-CM-002-V01	Carbon sink from bamboo afforestation	4 November 2013
AR-CM-003-V01	Carbon sink from forestry management	23 January 2014
AR-CM-004-V01	Greenhouse gas reduction and monitoring from sustainable management of grassland	23 January 2014
CM-096-V01	SF_6 emission reduction monitoring and calculation in GIS	15 April 2014
CM-097-V01	Utilisation of energy saving wire in new grid or old grid renovation	27 January 2015
CM-098-V01	Electric car recharging facility	27 January 2015
CM-099-V01	Small-scale eco-recovery for non-coal mine sites	27 January 2015
AR- CM-005-V01	Carbon sink from bamboo forest management	25 February 2016
CM-100-V01	Waste straw substituting timber for wood based panel production	25 February 2016
CM-101-V01	Greenhouse gas emission reduction baseline and monitoring for ready mixed concrete production	25 February 2016
CM-102-V01	UHV transmission system greenhouse gas emission reduction methodology	25 February 2016
CM-103-V01	Recovery of liquefied natural gas from coke oven gas	25 February 2016
CMS-079-V01	Greenhouse gas emission reduction from the utilisation of reactive power compensation device in distribution network	25 February 2016
CMS-080-V01	Building energy storage power station at existing or new renewable energy power plant	25 February 2016
CMS-081-V01	Ruminants farm stock emission reduction	2 June2016
CMS-082-V01	Livestock manure composting management greenhouse gas (GHG) emission reduction methodology	2 June2016
CM-104-V01	GHG emission reduction methodology for the preparation of low carbon ready mixed concrete and reduction of cement ratio by using recycled powder	2 June2016
CMS-083-V01	Methodology of GHG emission reduction and carbon sink from conservation tillage project	20 June 2016
CM-105-V01	GHG emission reduction methodology for bicycle sharing system project	22 July 2016
CMS-084-V01	GHG emission reduction methodology for the utilisation of radiation pyrolysis technology on household waste	26 August 2016
CMS-085-V01	Methodology for producing metallized pellet rotary hearth furnace using metallurgical solid waste treatment technology	26 August 2016

(*Continued*)

Table 6.1 New CCER methodologies approved by NDRC (*Continued*)

Methodology ID	*Name*	*Time of approval*
CMS-086-V01	Methodology for GHG emission reduction from efficiency improvement procedures of vehicles and fleets	26 August 2016
CM-106-V01	GHG reduction methodology for bio gas production and sales	26 August 2016
CM-107-V01	Biogas production and utilisation from manure management system	26 August 2016
CM-108-V01	GHG emission reduction methodology for new processing method of regenerative calcium carbide	18 November 2016
CM-109-V01	GHG emission reduction for direct reduction ironmaking technology of gas based shaft furnace	18 November 2016

Source: China Certified Emission Reduction Exchange Info-platform (2017).

In order to reduce carbon emissions while simultaneously improving urban air quality, the methodology for electric car recharging facility projects had also been developed (Methodology ID: CM-098-V01). However, it was not specified whether or not renewable energy was to be used. If using conventional mix electricity, electric cars are roughly four times cheaper to run than traditional petrol cars (calculated with medium conventional electricity price and petrol price). The cost of fuel for electric cars is much cheaper than cars running on petrol. However, the greenhouse gases emission of electric cars are not necessarily low if powered by conventional mix electricity because around 70% of the mix is generated from coal. Although electric cars may have the ability to reduce reliance on petrol and improve urban air quality, they do not reduce carbon emissions to a significant extent if coal-generated electricity is used (Pauliuk, Dhaniati & Müller 2012, p. 6). In order to achieve more significant emission reductions from the CCER electric car recharging facilities, it is recommended that electric car charging facilities should directly purchase electricity generated from a mix of CCER renewable energy projects. This can be achieved through administrative measures. Such regulation will not make electric cars less attractive because even the most expensive renewable energy is still three times cheaper than petrol.

Building energy efficiency is an area that has the potential to develop new low-cost abatement. In Table 6.1, the only CCER methodologies related to building energy efficiency are the ones which focus on the production of building materials such as Methodology CM-104-V01 and Methodology CM-101-V01. There is still no CCER methodology so far which targets the design efficiency of buildings. This book argues that the energy-efficient design of buildings can have a great impact on China's current and long-term climate change mitigation. This is because urban infrastructure, and especially office and residential buildings, usually have at least a 20–30 years of life span. Demolition and renovation are not likely to happen before this time; therefore, building energy efficiency has irreversible characteristics (Li & Colombier 2009, p. 2436). This book suggests creating CCER methodology on building design energy efficiency baseline projects, because energy-efficient designs can lead to many years of emission reduction in the future. Such projects should set the baseline requirement for energy efficiency factors including insulation, heating, cooling, and lighting. Energy efficiency ratings of new buildings are to be calculated based on annual energy consumption per square meter of floor area. Only building designs with high energy efficiency ratings should be considered potential candidates for this CCER methodology. It is also recommended to include car parks with electric car charging facilities as an essential component. This way, sizeable carbon emission reductions from the building sector can be guaranteed and emission reductions from the transport sector are also facilitated. It is recommended for the government to give priority approvals to residential and commercial buildings with the above low-carbon features.

China's newly developed low-cost emission abatement options are mostly small-scale, with consideration of sustainability. Such exploration should be continued in order to better facilitate the national ETS and truly embrace the essence of sustainability. To achieve this, sufficient financial support is needed. In the next section, the potential of CCER will be explored to determine whether it can play a greater role in China's national ETS. The importance of CCER finance for China's carbon off-setting program and the meaning of a successful CCER program to China's national ETS will be discussed.

CCER -- A greater role in China's ETS from the support of permits auction fund?

The emission reduction ability and environmental integrity of certain types of CCER projects are worthy of recognition such as renewable energy projects and the newly developed small-scale CCER projects mentioned above. However, it is predicted that CCER offset is limited at roughly 10% of the total national ETS allowance. The limit was set at 10% because companies may prefer to spend a smaller amount of money to buy CCERs rather than focusing on emission reductions from their own effort (21st Century Economic Report 2016, n.p.). However, this book argues that the proportion of CCERs should be made more flexible: a bigger proportion of CCER in China's carbon market will not harm the climate integrity of China's nation ETS if the carbon emission allowance is not over-allocated. To the contrary, bigger proportion of CCERs may bring economic efficiency. This is because many of China's industries included in the national ETS have reached fairly high-energy efficiency which makes further efficiency improvement extremely expensive. In China's power sector, for example, the majority of the newly built large-scale coal-fired power plants were already quite efficient. The only way to make them even more efficient is to upgrade to more advanced combustion technologies such as the ultra super critical or integrated gasification combined cycle, which can moderately improve the combustion efficiency (Shen 2015, p. 344). However, these advanced technologies are so expensive that the installation will always be more costly than paying for the fines for failing compliance. Under these circumstances, companies would prefer to pay the penalty for failing compliance rather than install the advanced equipment. As such, the readily available offsetting option is necessary. This is where abundant and quality CCERs is needed. For industries with either limited technological upgrade potential or inability to comply just by reducing production, offsetting becomes the only solution for achieving compliance under a cap and trade system.

There are concerns that a bigger proportion of CCERs may inflame the market causing the carbon price drop. However, this book argues that such a situation will not occur if the original carbon emission permits are not over-allocated. Even better, the original permits should be auctioned off rather than directly distributed to industries. This is because, first, industries will conduct their own analysis to decide a more accurate amount of permits they need. Just as Benz and Trück (2009) put it:

> The CO_2 allowance price is determined directly by market scarcity induced by the current demand and supply at the carbon market. Notably, firms by themselves are able to control market scarcity and hence the market price by their CO_2 abatement decisions. (p. 9)

Second, the government will obtain a financial gain from the action which can be fed back into the carbon market. For instance, the fund can be utilised to provide special loans to quality CCER projects. As mentioned in Chapter 4, the finance for small-scale CDM projects was at a bottleneck. Small CCER projects face the same problem today.

According to the interview Shen (2015, p. 347) conducted within the Chinese financial sector, few mature financing products had been developed to support the development of CCER projects due to the difficulty in the assessment of profit outlook of CCER projects. The government can combat this deficiency by setting up a special CCER fund from the financial gain of permit auction. It is already recognised by many that China's carbon market follows a top-down fashion in which the state weighs more than the market (Goron & Cassisa 2017 Lo & Yu 2015; Shen 2015). Therefore, it does not hurt for the government to do a bit more to perfect the whole system. The Chinese government have already built up the framework of the national

ETS, it should start to add detailed policies and guidance such as providing sufficient financial support and other services to relevant stakeholders, and in this case, to quality CCER projects. The purpose of this is to show banks and other financial institutions that financing CCER projects can be profitable. A successful outcome of the special CCER fund will have the potential to lead banks and other financial institutions to start developing CCER financial products and strengthen carbon finance in the offsetting mechanism.

So far, China inherited CDM as a domestic carbon off-setting mechanism; however, the current policy and regulation of national CCER are still not in place. This is what the Chinese government should urgently work on. Nonetheless, relevant institutions and climate consultancies were indeed very active in developing new low-cost abatement options. Majority of the CCER methodologies approved by NDRC follow the principle of sustainable development and can justify their additionality. However, contentious methodologies such as the ones for large-scale hydro-electricity projects and industrial gas projects should be excluded from the CCER methodology list in order to safeguard the environmental integrity of China's CCER program and the national ETS. Lack of financial support for CCER projects is a major drawback that restrains the potential of CCER development. This book suggests the government should set up examples for the financial sector by financing quality small-scale CCER projects from the financial gain of permits auctioning. These are the implications this chapter drew out to address the issues associated with China's domestic carbon offsetting mechanism and carbon markets. In the next section, it will focus on addressing the issues raised in the analysis of China's domestic energy saving and emission reduction policy – ECERS.

Update of ECERS in the 13th Five-Year Plan and implications for problems raised in Chapter 5

Chapter 5 of this book analysed China's major domestic climate-related policy – the ECERS during the 11th and 12th Five-Year Plan period. It focused on the policy and implementation of ECERS in the energy production sector and the Energy Performance Contracting (EPC) for energy end users. The analysis identified that the major issue with China's ECERS was neither a scale problem nor a speed problem, but the integrity and efficiency problems in the implementation process. In total, four major weak aspects of ECERS need to be carefully addressed including laid-off workers' welfare, fraudulent claims of cash rewards, the restricted production of renewable energy, and the potential of EPC. This section will follow up policy updates for addressing these issues during the 13th Five-Year Plan. For issues that have not been dealt with, some implications will be provided.

Dealing with laid-off workers' welfare

The effectiveness of ECERS was undeniable in closing down heavy polluting small coal-fired power plants. However, the compensation and welfare for laid-off workers and private owners should be handled with more care. The phasing out of outdated production capacity during the 11th and the 12th Five-Year Plans can be seen as the second wave of worker redundancy since China's reform and opening up in the late 1970s. The first wave of worker redundancy was the result of the SOE reform during the late 1990s. The second wave was very different from the first wave in terms of scale and workers' perceptions. In the 1990s, greater number of workers from SOEs were laid-off and workers showed greater resistance. This was mainly because workers from SOEs were so used to the planned economy and were shocked by the concept of the market. Lee (2000, p. 271) described the workers reaction as the "revenge of history" as during Mao's era workers' income were guaranteed and interests protected. The workers' perception differed slightly towards redundancy during the recent phasing out of outdated production capacity from previous SOE laid-off workers. Although upset, they were less resistant to the idea of seeking new employment in the labour market or becoming self-employed. This was due to a few reasons. First of all, not all closed down small coal-fired power plants were SOEs, some of them were rural collective owned or privately owned. Therefore, the workers were not as dependent on their employers as some of the previous SOE laid-off workers. Second, some of the closed down small coal-fired power plants were already under financial pressure and heavily in debt. Workers' salaries were kept at the rate of the minimum wage or even suffered from delayed payment. Therefore, there was a motive for workers to seek new and better employment opportunities. Third, recent laid-off workers were more used to the concept of the market after the three decades of market reform. The fast-developing and popular e-commerce sector provides vast opportunities for small business which require little upfront investment (Wang, Cooke & Lin 2016). Under these new circumstances, reasonable compensation and practical retraining programs appear to be more important.

How to make unemployment insurance incentive based social welfare

The unemployment insurance (UI) scheme was first introduced to China in 1986 to deal with the emerging unemployment problem. The original scheme only covered SOE employees where employers pay 1% of the employees' salary (Leung 1995, p. 144). The UI broadened its coverage over the years. Since 1999, UI became a mandated social security fund covering all urban workers with formal work contracts to which the employers contribute 2% of their payroll and employees contribute 1% of their salary (Vodopivec & Tong 2008, p. 14). The fund provides an allowance to the qualified workers[1] for up to 24 months after their redundancy. The UI allowance should not exceed the regional minimum wage and may range between 40% and 80% of the worker's previous 12 months average salary depending on localities (The Central People's Government of PRC 2005, n.p.). There are new updates for UI since 2018. According to Ministry of Human Resources and Social Security (MOHRSS) of PRC, the UI premium rate is to be lowered to 1% across the country, while the UI payment will increase to 90% of the local minimum wage (MOHRSS 2018). This move will help to provide relief for the unemployed.

In the late 1990s, the UI fund was once severely strained due to the large number of laid-off SOEs workers. Currently, the situation is quite the opposite. The UI fund in many localities today is in large surplus. For instance, the UI fund in Wu'an city was 47,910,000 yuan (approximately US$8,000,000) in 2013, while the UI claim was only 4,060,000 yuan (approximately US$680,000) during the same period (Xinhua News 2014, n.p.). At the moment, the UI is limited to providing a minimum allowance to maintain workers' basic living standards after redundancy. Apart from this, it does not provide other forms of assistance to the unemployed. It is a cushioning payment to guarantee survival but is not very effective at encouraging workers to seek reemployment. This single-purpose payment is also the main cause of the large surplus in the UI fund.

This book recommends doubling the current UI allowance rate but reducing the maximum UI allowance duration to 12 months, meanwhile providing low or even zero interest mini loans to eligible candidates who wish to start a small business or become self-employed. The amount of the loan can be capped at two to three times of sum payment of 12 months allowance depending on different UI funds condition. It is predicted such an adjustment will provide incentive for the unemployed to actively seek new opportunities.

The maximum 24-month duration of UI allowance is quite lengthy compared with 12 months in many European countries (Vodopivec & Tong 2008, p. 31). The lengthy social security support period may weaken workers' will and enthusiasm for finding new jobs, and may even reach a status that the workers get used to living on minimum social welfare (Farber & Valletta 2015, p. 909). Therefore, the duration should be reduced. The allowance rate, meanwhile, should be doubled. This way, the total amount of cash allowance is unchanged. There are critiques of the generosity of some European unemployment benefits which create disincentives for reemployment (Nickell 1997). However, China's UI payments are much lower than that of the EU and only satisfy basic survival. Thus, the situation in EU does not apply to China. Doubling the current Chinese UI rate will allow a moderate benefit payment. The recipients can gain access to more resources such as expenditures for transport, accommodation, internet, and communication during the process of seeking new employment opportunities (Tatsiramos & Ours 2014, p. 289).

In order to maximise workers' enthusiasm in seeking new employment, this book also suggests offering micro-loans with UI funds for the eligible candidates who wish to start a small business. For those who have successfully completed retraining, the UI loan can provide unemployed workers a wide range of opportunities in the service industries. According to Kumar, Hossain and Gope's (2015) experience in Bangladesh, micro-loans can have a range of positive social and economic benefits. The UI loan is still a loan which requires repayments, which will ensure the health of the UI fund as there will be cash flow back to the fund. However, UI loans only require repayment when an applicant's new income reaches a certain level. This is very similar to Australian university students' Higher Education Contribution Scheme (HECS). Therefore, for applicants, UI loans are risk free. Some may disagree with this proposal by arguing that the UI is not designed to provide loans. It is the banks' function and responsibility to do so. However, a large proportion of the unemployed workers who rely on the UI allowance do not own significant assets, so it may be difficult for them to pursue a loan from the bank. Even if they have the ability to acquire a loan from the bank they may be reluctant to do so due to worries about risks and the consequences caused by failed repayments. Therefore, UI loans provide great resources for workers' reemployment. Apart from resources, new skills are also important for workers who seek reemployment. Some suggestions for workers' re-training programs will be provided in the following section.

Scholarship entitlement for retraining program

There are several problems with current workers' retaining programs. First, workers have few choices in the selection of new skills. Second, the topics offered appear unpopular in the labour market. Third, the re-training programs are usually organised by the workers' previous employers such as some SOEs or the local welfare community which are under-resourced and the training schedule is inflexible (Ma 2016, p. 249; Wu 2009, p. 59). In response to these weaknesses, this book suggests that the government should liaise with local qualified private training institutions and establish a scholarship system for the unemployed. The initial step that relevant departments of the municipal government should undertake is to research and identify several high-quality and well-known private training institutions in the locality. And then each qualified unemployed worker is granted a scholarship which can be used in any of the appointed training schools. When the worker completes his/her enrolment, the UI fund will transfer the tuition fees to the relevant training institution. Being private and competitive, each of these training institutions should have already been providing skills training with specialty that is in demand from the local labour market. Therefore, the students will not only have more choices, but also higher quality training, which will give them confidence and in-demand skills for reemployment or self-employment. Moreover, the combination of a scholarship system and the private training school also gives students flexibility to choose classes suitable for their own time schedule. This has some similarities to the assistance provided to laid-off workers in the automotive industry in Australia. For example, the Victorian state government, federal government, and Ford provided AU$135 million for grants, programs, and training for automotive industry workers who lost their jobs as a result of the closure of the Ford factories in Broadmedows and Geelong. Some of these focused on seeking similar opportunities where the employed can use their existing skills, while others focused on equipping them with new skills (Toscano 2016, n.p.).

The late 1990s SOE reform had transferred large amount of SOE workers to the private sectors with many of them relocated to the light industry in the Southeast coastal regions. This process is now almost completed. The light manufacturing industry targeting the international market is already saturated and has the tendency to shrink due to the slowdown of the global economy. Therefore, the newly unemployed from the process of eliminating outdated production capacity are most likely to obtain new employment opportunities within the service industry. Therefore, it is essential to utilise all quality resources to provide training to workers targeting the future market and growth in the service industry. Such an initiative has been proved effective in Taiyuan city, where free retraining has been jointly provided by the UI centre and private training organisations (Wang 2017, n.p.). Furthermore, the blooming e-commerce will also bring many opportunities for small business and self-employment. For instance, according to Wang, Cooke and Lin (2016, p. 292), China has the biggest population and proportion of informal employment compared with formal employment. Self-employment through online business is one example among all forms of informal employment. Liu, a laid-off worker from Xiangyang city, started her own online store selling local produce. Her business has been growing due to her hard work and continuous study of e-commerce marketing. As her business expands, she plans to employ some of her fellow laid-off workers for her business (Chen 2015, n.p.). This further indicates the feasibility of e-commerce as a channel for reemployment.

Rethinking of cash rewards as an incentive for energy saving and emission reduction

By the end of the analysis period, 2015, to be exact, the Chinese government decided to stop handing out direct cash rewards for SOEs and Energy Service Companies (ESCOs), instead, it provided only tax benefits for eligible corporates' energy saving and emission reduction achievements. This is to say, the government, as shown in Chapter 5, had already realised the drawbacks of cash reward as an incentive. The cash rewards were set up with goodwill but it ended up as bait for many cases of corruption and fraudulent claims. The fraudulent claims did not reduce in the 13th Five-Year Plan. Henan, Yunnan, Guangxi, and part of Guangdong were all accused of "superficial", "fake", or "perfunctory" rectifications according to the latest MEE review (Stanway 2018, n.p.).

According to the analysis in Chapter 5, the opportunity for fraudulent claims was embedded in the universal use of the administrative approach in managing energy conservation and emission reduction for enterprises of all types of ownerships. Such management was effective in dealing with large SOEs under direct supervision of the central or provincial government, but was not so for non-SOEs especially those in remote counties and towns. In such places, direct administrative supervision was far from reach, auditing from powerful administrative departments was minimal, and the third-party auditing was next to none. As a result, there had been cases of fraudulent cash claims for unachieved energy saving and emission reductions. On one hand, it indicated that cash reward was trouble-prone without a precise auditing system; on the other hand, it reflected mono-policy approach, the administrative approach, was not sufficient in managing all types of enterprises.

Avoiding direct cash rewards and adopting tax incentives can effectively avoid fraudulent claims. For future policy development on energy conservation and emission reduction, direct cash handouts should be avoided especially when only administrative approach is used and auditing is lacking. Also, it is important to deal with SOEs and non-SOEs with different approaches. Administrative approach is suitable for managing SOEs because the top executives of SOEs are appointed by the state, who are, in nature, equal to government officials. How well they implement central government's policy is closely related to their promotions and future status. However, non-SOEs are market entities. A pure administrative approach will not be as effective. This is because they are profit-driven and will find ways to get away with the unfavourable administrative orders especially when there is no effective auditing body around. Therefore, for non-SOEs, a combination of market approach, legal tools, and possibly civil society watchdogs is recommended. The national ETS scheme is hopeful to offer such a medium in the foreseeable future. It can provide systematic monitoring and auditing for corporates' emissions, and eventually bring in laws and regulations which clearly define the punishment for corporates who fail to comply with the set emission target. The effectiveness of the scheme is yet to be seen, but it does point out a new direction for China's management of national greenhouse gas emission for both SOEs and non-SOEs.

The foundation of a national ETS should be a well-established auditing system which collects reliable data of corporates' emissions through constant monitoring. This process will overcome the auditing weakness in China's traditional administrative system, which basically relies on cross auditing of written reports by randomly selected auditing departments of different regions. Although the third-party auditing had been introduced recently, the qualified third-party auditing agencies were limited. The roll out of national ETS requires the strengthening of national auditing system which would make it difficult to falsely claim emission reduction quantity.

Furthermore, putting a price on carbon increases the competitiveness of renewable energy and makes inefficient coal power less attractive. In response, major energy SOEs may reconsider the proportion of renewable energies in their portfolio. Instead of passively keeping a minimum renewable energy for the sake of showing the central government, they have followed the instruction; they may voluntarily increase the renewable energy proportion. For inefficient non-SOEs, market competition and high carbon price may put pressure on their operation and survival and may eventually phase them out of the market. This result may be easier for them to accept than being forced to shut down by direct administrative order when they are still making profit.

How to truly and effectively utilise renewable energy

During the 11th and 12th Five-Year Plans, China's renewable energy capacity expanded in significant speed and scale. China continued to lead the renewable energy investment in the world after the 2015 Paris accord. In 2016, China invested nearly US$88 billion in renewable energy, which was over 30% more than the United States. In early 2017, China committed another US$360 billion to its renewable energy portfolio as a mean to combat global climate change (Shankleman 2017, n.p.). However, there is one thing that needs to be addressed: China did not utilise its full renewable energies capacity. In fact, a significant proportion of China's renewable energy was wasted.

There has not been significant improvement dealing with wasted renewable energy during the 13th Five-Year Plan. According to the Greenpeace, wind power wasted in China increased to 17% in 2017. The unconnected wind power alone was enough to power Beijing for a whole year in 2016 (Stanway 2017, n.p.). China's grid construction was indeed a little lagged behind the speed of wind and solar capacity. However, according to the analysis in Chapter 5, grid restriction was not the biggest obstacle of China's renewable energy wastage. After all, the wind and solar energy which was not connected to grid was only a small proportion compared with the total wasted renewable energy. A large proportion of the wasted solar and wind power was in fact connected to grid but the production was restricted by administrative orders and dirty local politics. Many private renewable energy corporate received unfair treatment due to this.

Now the most important issue that needs to be addressed is how to restrict provincial and municipal governments' outrageous power over the decision of which power plant can operate at full capacity and which cannot. The central government's planning on the development of solar and wind power had been consistent with the development of China electricity transmission grid. It was some provincial governments that blindly installed extra renewable energy capacity beyond the central government planning for attracting relevant manufacturing industry such as wind turbines and solar panels so as to raise tax revenue; and again, for raising tax, they approved more coal power capacity when the coal mining lobbyists promised more benefits. When both renewable and coal-fired power plants were in operation, they purposely gave priority to coal-fired power plants to sustain the coal-mining industry in their administrative region, and restricted the generation of renewable energies. Such behaviour seriously violated the *Renewable Energy Law*. In order to constrain provincial and municipal governments' decision-making power in energy planning and generation, this book proposes several strategies.

First of all, it is suggested to appoint a top official (i.e. secretary of provincial party committee; secretary of municipal party committee) within the provincial and municipal government to be responsible for local renewable energy development strictly according to the renewable energy law. Such initiative has been taken by mayor of Beijing, Wang Anshun, in dealing with air pollution. Wang Anshun signed up a duty pledge with the central government in the effective management of air pollution. During his current term of office, he shut down more than 2000 polluting companies, converted coal-powered central heating to natural gas heating, and controlled traffic emission through passenger car on-road schedule regulations (Liu 2014, n.p.). These actions led by the mayor relieved air pollution in Beijing. In the supervision for renewable energies, the *Renewable Energy Law* clearly stated that renewable energy generation should be given top priority when total local energy generation exceed market demand. The responsible official should respond directly to the central or provincial government and should be questioned for any amount of wasted renewable energy. This proposed method is similar to China's Target Responsibility System (TRS) which was effective in many tough reforms, for instance, the shut-down of heavy polluting small coal-fired power plants as mentioned in Chapter 5. How well the official manages local renewable energy will have an impact to his or her future promotion. Because the appointed officials are top leaders in the provincial and municipal governments, it is unlikely that their authorities are challenged even if other members of the government are in favour of coal power.

Second, it should be made illegal for administrative authority to intervene with renewable energy corporates' production. Provincial and municipal governments have no right to order renewable energy corporates within their jurisdiction operate at certain percentage of their capacity; nor are they allowed to threaten or punish the renewable energy corporates. If there are renewable energy corporates who do have unlawful operations, the government should pass on the case to the legal system to judge and sentence.

Third, provincial and municipal governments should be punished for wasting renewable energy in their jurisdiction. In China's carbon trading pilots, corporates failing to comply with their emission reduction targets were punished. So why the government who purposely wasted renewable energy and prevented emission reduction should be left unchecked? This book suggests deducting whatever loss caused by the wasting renewable energy from relevant provincial and municipal governments' tax revenue. The deduction should include both the cash value of the wasted renewable energy and the amount of CCERs they represent at market price. The money is collected by the relevant department of the higher government to put into a special fund for carbon offset. Tough penalties are necessary and will be effective for eliminating future wastage of renewable energy. It is suggested the penalties for other forms of violation of the *Renewable Energy Law* should also be tougher and clearer. The *Environmental Protection Law* has already been made stronger and penalties for various breaches increased as discussed in Chapter 3, so it can be assumed that it will continue to do so in future, and incorporate the suggestions made here.

Apart from restraining provincial and municipal governments' power, how to increase the strength and resilience of renewable energy corporates also deserve discussion. With the establishment of national carbon market and the role out of national ETS, the renewable energy industry will enjoy a more favourable condition. The added price on carbon emission will make renewable energy relatively cheaper and therefore make it overall more competitive. Several issues need to be considered for future policies within the renewable energy sector such as creating a fair market environment, providing extra financial service to private investors, and free of discriminations against CCERs generated from small private renewable energy enterprises. These issues may take time to be fully addressed, and some policies may even come out after problems emerge in national ETS. However, for current utilisation of renewable energy, this book suggests two possible solutions to prevent wastage.

First of all, as the development of electric car recharge facilities had been included into China's CCER project category, this book suggest to give priority approvals to real estate development with such facilities and make renewable energy a mandate fuel source for electric cars. According to Interviewee 14, the biggest obstacle that prevents the popularity of electric cars was the lack of recharging facility and space. Many of China's older apartment buildings are without car parks. He had seen electric car owners dropped an extremely long cord from the sixth floor window to the ground to recharge the car. If the car owner cannot find an off-street parking space near his window, it would create hassle. Therefore, this book suggests the new real estate development to be located on the outskirts of mega cities or as satellite cities of mega cities with smart urban planning and sufficient supporting facility. Business and employees are encouraged to move to the new developments where rent and housing price are lower. This will not only relieve the traffic condition in mega cities, reducing air pollution, but will also demonstrate the possibility of universal adaptation of electric cars and set up example for future low carbon cities. In doing so, China's future transport sector can consume a large proportion of China's renewable energies, and the demand of renewable energy can be sustained if electric car ownership becomes sizeable.

Second, it is suggested that China's northern provinces should consider absorbing wind energies within the region rather than sole relying on exportation. The pattern of wind energy peaks during winter months, and northern provinces consume large amount of energy for central heating. At the moment, winter central heating is largely powered by coal. To be exact, an average of 130 million tons of standard coal are used annually, which roughly equals to 433 billion kWh electricity (China IRN 2014, n.p.). China's wind power production in the entire 2016 was only 241 billion kWh (China bgao 2017, n.p.), about half the demand of central heating. This is to say, wind power can definitely replace part of coal as an alternative energy source for winter central heating. It is feasible to be consumed within the northern region of China, and the lack of a north-south transmission grid for exportation will no longer be an excuse for wastage. If priority is truly given to renewable energy as required in the Renewable Energy Law, China in theory should have wasted no wind and solar energy.

Innovations for EPC

EPC was an important market mechanism that the Chinese government introduced to ECERS. The purpose of EPC was to create a win-win solution for both the ESCOs and their clients. The ESCOs would make a profit while their clients could achieve energy saving and emission reduction at zero cost. China's EPC was growing at a relatively fast speed. However, it also encountered some problems, namely the lack of trust among EPC stakeholders, lack of formal finance, and difficulty in finding suitable projects. These problems were not isolated. Rather, they were interconnected with different EPC stakeholders. Therefore, this book focuses on innovations that sooth the relationship among EPC stakeholders rather than tackling each individual barrier of technology, market, and finance. Just as Kostka and Shin (2013) put it, "in the context of contemporary China ... trust and networks are critical features shaping current outcomes" (p. 750).

Improving trust among EPC stakeholders

Chapter 5 identified that the lack of trust among EPC stakeholders was one of the biggest obstacles that restrained the development of EPC in China. Taylor (2009, p. 34) who conducted the World Bank EPC programs in multiple developing countries pointed out that a successful EPC program required sound institutional development which could effectively bring together financial providers, ESCOs, technical experts, and clients. In order to strengthen the relationship among EPC stakeholders, this book believes it is essential to introduce a clear and sound accreditation system for the ESCO industry; meanwhile, ESCOs should also conduct more precise risk analysis when dealing with potential energy conservation projects.

The quality of Chinese ESCOs vary greatly. Kostka and Shin (2013, p. 749) conducted an extensive series of interviews with Chinese EPC industry insiders who described Chinese ESCOs as a motley collection of enterprises. The majority of Chinese ESCOs were small with limited technological and financial capacity which appeared difficult to build up trust and reputation among local businesses. Many of them, instead of designing the full energy conservation solution and service for their clients, mainly aimed at selling the products they manufacture. Some ESCOs preferred to be called so solely for the purpose of claiming financial and tax benefit. Therefore, there is an urgent need for setting up a national standard accreditation system for ESCOs in order to guarantee the consistent industry credibility. The unified accreditation system may consider multiple criteria for ESCOs such as technology assessment, human resource assessment, service assessment, finance assessment, credit assessment, and periodical auditing. Such accreditation system can be jointly developed by ESCO Committee of China Energy Conservation Association (EMCA), relevant NDRC research institutions, and experts and insiders from the EPC industry, and it should be applied to all ESCOs in China. The universal accreditation system will provide a clearer display of ESCOs' expertise. The clearer rating is expected to display a clearer guidance for potential energy conservation clients when choosing an ESCO. ESCOs with sound service quality, exceptional expertise, rich experience, and good reputations are likely to obtain high ratings from the accreditation system, and therefore, can build up confidence within their potential customers. In this way, the quality and trustworthiness from the ESCO side of the relationship can be strengthened.

ESCOs should also improve their own ability in exploring and identifying quality energy conservation projects. ESCOs need to apply a more comprehensive risk analysis to their potential clients and projects. In doing so, they need to employ personnel with expertise in risk analysis. If such employment is not considered, they would need to hire professional consultancies to conduct the analysis. In a word, the risk analysis procedure should not be left out, but be conducted in a comprehensive manner to ensure sound return of their investment. Another way to reduce risk for ESCOs is to get the insurance industry involved. If certain insurance company can develop products catered for EPC projects, it would be extra layer of protection for ESCOs' investment.

Increasing the involvement of formal finance

The financial providers added a third dimension in China's EPC stakeholder relationship. Currently, the majority of EPC finance are sourced from less formal financial channels such as shareholders and strategic partners. The involvement of formal finance is limited apart from the World Bank and Chinese government appointed bank loans (Taylor 2009, p. 40). Again, this is a trust issue. Banks are not familiar with the operation of EPC projects, nor are they convinced by the methods used in evaluating EPC project performance (Kostka & Shin 2013, p. 757). To turn around this situation, government administrative pressure as proposed by Gan (2009, p. 1701) alone will not work. This is because banks also need to consider their own risks and make profit. In order to fundamentally improve the involvement of formal finance, this book suggests building up the relationship between banks and ESCOs through solid data, categorisation, and transparency. The ESCO industry should actively promote itself to its potential clients and financial providers. For instance, EMCA can initiate yearly industry report specially targeting potential financial providers. The report should contain detailed data of all EPC cases including hosting ESCOs information, methodology and technology adopted, finance, profit, etc. The report should also categorise EPC projects according to different criteria such as technology, methodology, and scale. Such a publication will give banks a clearer idea of current EPC market and leave them to explore the potential in EPC financing. To better assist the promotion of EPC, the report should be made public and easily accessed online. Potential financial providers, ESCOs, and potential EPC clients should also be informed and updated by invitations to conferences catered for EPC industry and market. This provides convenience for financial organisations to conduct risk analysis and obtain overall data of the EPC industry (Jiang & Li 2013, n.p.).

The finance bottle neck that the ESCO industry faced was largely branched from its own status of being new, small scale, and unknown. Banks lacked motives to approach such kind of new and less significant business form. Therefore, the ESCO industry needs to actively promote itself and create opportunities for potential financiers and clients to know about EPC. Displaying accurate and rationally organised data is one possible way to achieve this and acquire trust. China has a big market for EPC due to its resource intensive and low energy efficient production model (Li 2012, p. 121). There are numerous successful cases in which ESCOs professionally managed energy conservation and created win-win solutions. Therefore, the key for broadening up finance for ESCOs is to let financial providers know that EPC projects are effective and there is profit be made in EPC market.

Establishing info-exchange platform for ESCOs and potential clients

Apart from the trust issue among EPC stakeholders, there are also practical problems. For instance, ESCOs do not have diversified channels in identifying potential clients and projects; and at the same time clients who wish to conduct EPC projects have limited information about local ESCOs. This makes business difficult in the initial stage. In response to this problem, this book suggests to establish a range of EPC info-exchange platforms at both local and national levels. Such info-exchange platforms are web-based and open to all ESCOs and EPC clients. ESCOs can display their service, specialty, and showcase projects to potential clients. While EPC clients can also publish their energy conservation project requirements on the platform to attract ESCOs with required expertise. Such info-exchange platforms are capable of creating better matches for ESCOs and clients so that to improve business efficiency. Although currently the EMCA fulfils such a role, it does so more on a national and policy level. On the local level, such services are absent. There are local initiatives for improving the communication between ESCOs and their potential clients. For example, in Shanghai, a symposium was hosted for 26 ESCOs and 100 companies which wished to improve energy efficiency. The symposium provided a face-to-face opportunity for ESCOs and their potential clients to get to know each other, where they could find the perfect match for their needs (Lu 2015, n.p.). However, such an event is not regular. If such an event can be delivered from an online network, ESCOs and their potential clients would be able to access the information they need anytime. Thus, the efficiency of ESCOs can be improved.

Also, many high-quality and reputable ESCOs are located in big cities, while many companies who require energy conservation through EPC are often located in less central places where it is hard to access good ESCOs. This book suggests a mobile project manager approach for large quality ESCOs. ESCOs can appoint projects managers who are willing to travel to specific locations to conduct initial assessment of an EPC project and also conduct a series of follow up services. Mobile project managers are payed a decent commission for successful EPC projects they assist to implement. This initiative can be a good way of introducing high-quality EPC management to relatively remote areas where less efficient energy intensive industries are concentrated.

Encouraging well-established ESCOs to explore opportunities in CCER projects

China's ESCO industry is still young. However, high-quality ESCOs do exist. For instance, China's top ten ESCOs ranked by Qianzhan Industrial Research Institution (2015, n.p.) are all of great quality. It is recommendable for quality ESCOs to explore CCER opportunities. ESCOs can investigate into energy efficiency projects which fit CCER methodologies. The targeted CCER projects are not limited to large-scale energy efficiency improvement projects. In fact, ESCOs with sufficient resources can bundle and manage a range of small and similar CCER energy-efficient projects. Instead of receiving cash rewards from the government, ESCOs now can earn CCERs on the carbon market. China's existing high-quality ESCOs have the potential and capacity to cover China's energy efficiency improvement CCER projects. The two market mechanisms CCER and EPC are not contradictory. In fact, CCER can include EPC as a major force for the development of energy conservation projects. It is believed such EPC projects will be more formal and precise when measured by the more sophisticated CCER operational and auditing procedures.

EPC will have its role and great potential in China's national carbon market. However, to improve the quality of Chinese ESCOs and to enhance business between ESCOs and their clients, the trust issues need to be solved. This section has suggested a unified national accreditation system for ESCOs, adopting more precise risk management strategies and possibly introducing insurance products. To remove the financial bottleneck that many ESCOs encountered, it is recommended for ESCOs to produce annual industrial report specially targeting potential formal financial providers. In order to overcome practical difficulties, this book also suggests to establish info-exchange platforms for ESCOs and their clients. High-quality ESCOs are also encouraged to explore business opportunities in CCER projects to keep utilising their expertise in energy conservation and energy efficiency improvement.

Summary

This chapter followed up updates in China's climate-related energy policies during the 13th Five-Year Plan (2016–2020). It investigated what policy weaknesses raised in Chapter 3–5 had been addressed by these updates, and what had not. It found that the most significant update in the current Five-Year Plan was in the government actor. The MEP was renamed the MEE and took over the management of the DCC from the NDRC. It is expected that such a move would bring positive outcomes in China's future climate change governance. Another significant update was the implementation of the national ETS and the utilisation of CCER as a carbon offsetting mechanism in the scheme. For issues that had not been addressed, this chapter discussed a range of implications. These included (1) a WG approach to address conflicts and competition among climate-related ministries, (2) university education and other academic certification to the public to boost personnel with climate expertise at provincial and local government level, (3) protecting renewable energies through the promotion of green-energy-only sectors such as electric car charging facilities and winter heating in China's northern provinces, (4) medium and small renewable energy investors are recommended for extra financial support generated from the fund of auctioning of carbon credits, (5) China's ENGOs can be strengthened by improving environmental laws and an easier legal process for environmental lawsuits, (6) an innovation between EPBs and ENGOs can be carried out by providing a subcontracting system to deal with polluters, (7) ENGOs should create more channels for the general public to participate in local environmental discussions and to report suspicious polluters, and (8) the general public should also be provided with more diverse options for their energy sources such as renewable energy options.

For domestic carbon offsetting program, (1) the Chinese government should work on a comprehensive national CCER policy, (2) contentious methodologies such as those for large-scale hydro-electricity projects and industrial gas projects should be excluded from the CCER methodology list to safeguard the additionality and environmental integrity of China's CCER program and the national ETS (3) the government should also encourage domestic financial institutions' involvement in low-carbon finance by setting up positive examples of financing quality small-scale CCER projects from the financial gain of permits auctioning.

For ECERS, (1) laid-off workers' welfare should be guaranteed through the reform of the UI, (2) cash rewards for emission reduction should be replaced by tax breaks in order to prevent fraudulent claims, (3) provincial and local government officials who forced renewable energy companies to operate at a restricted capacity through administrative order should be punished, (4) the loss of the affected companies should be compensated from the government's tax revenue, (5) central heating in northern provinces should utilise wind power as a priority, (6) electric cars should utilise renewable energy and the government should also give priorities to real estate development with electric car charging facilities, (7) building up a comprehensive information exchange platform for EPC to improve communication and trust among EPC stakeholders, and (8) encouraging quality EPC projects to apply for CCERs.

List of interviewees

Interviewee 1, Officer of NDRC

Interviewee 3, Officer of MEP

Interviewee 14, Professor from the School of International Studies, CASS

Note

1. To be eligible for the UI payment, workers need to meet three conditions. (1) Having participated in a UI scheme for at least 12 months before redundancy; (2) The redundancy was involuntary; (3) Registering for unemployment status within 60 days after the termination of the work contract.

References

21st Century Economic Report 2016, *2015 nian qian de CCER buneng yongyu quanguo tanshichang? Qihousi zheme shuo (CCERs before 2016 cannot be used in national carbon market? Department of Climate Change said so)*, viewed 11 February 2017, <http://www.tanpaifang.com/CCER/201604/2052469.html>.

Anguelovski, I & Carmin, J 2011, 'Something borrowed, everything new: innovation and institutionalization in urban climate governance', *Current Opinion in Environmental Sustainability*, vol. 3, no. 3, pp. 169–175.

Australian Public Service Commission 2004, *Connecting government: Whole of government responses to Australia's priority challenges*, viewed 04 January 2016, <http://www.apsc.gov.au/publications-and-media/archive/publications-archive/connecting-government/challenge>.

BCCRC 2012, *Jidi fazhan jinqi ji yuanqi guihua (Center's immediate and long-term development plans)*, viewed 19 November 2018, <http://bccrc.ucea.edu.cn/jdfzjqjyqgh/index.htm>.

Benz, E & Trück, S 2009, 'Modeling the price dynamics of CO 2 emission allowances', *Energy Economics*, vol. 31, no. 1, pp. 4–15.

Cendrowski, S 2014, *Why China's rich are leaving*, viewed 10 October 2017, <http://fortune.com/2014/06/05/china-rich-immigration/>.

Chen, CD 2015, *Xiagang nvgong kai wangdian (Femail laid-off worker starts an online store)*, viewed 7 July 2017, <http://www.xf.gov.cn/news/xyxw/xyyw/201503/t20150303_595789.shtml>.

China bgao 2017, *2016 nian quanguo fengdian fadianliang zhan quanbu fadianliang 4% (2016 wind power generation reached 4% of national total energy generation)*, viewed 1 May 2017, <http://www.chinabgao.com/stat/stats/81747.html>.

China Certified Emission Reduction Exchange Info-Platform, 2017, *Fafaxue (Methodology)*, viewed 14 February 2017, <http://cdm.ccchina.gov.cn/zylist.aspx?clmId=162>.

China IRN 2014, *Quanguo gongre cainuan haoneng quannian yue 1.3 yi du biaomei (National central heating consumes approximately 130 million tons of standard coal)*, viewed 1 May 2017, <http://www.chinairn.com/news/20140704/152405348.shtml>.

Christensen, T & Lægreid, P 2007, 'The whole-of-government approach to public sector reform', *Public Administration Review*, vol. 67, no. 6, pp. 1059–1066.

Council of Australian Governments (COAG) 2017, Role of COAG Councils, viewed 10 October 2017, <https://www.coag.gov.au/coag-councils>.

Department of Agriculture and Water Resources, Australian Government 2017, 'Council of Australian Governments water reform', viewed 13 October 2017, <http://www.agriculture.gov.au/water/policy/coag>.

Department of Climate Change (DCC), NDRC 2012, *Guojia fazhan gaigewei guanjyu yinfa <wenshi qiti ziyuan jiaoyi ganli zanxing banfa> de tongzhi (NDRC releases the temporary management method for CCER projects)*, viewed 15 January 2017, <http://qhs.ndrc.gov.cn/zcfg/201206/t20120621_487133.html>.

Farber, HS & Valletta, RG 2015, 'Do extended unemployment benefits lengthen unemployment spells? Evidence from recent cycles in the US labor market', *Journal of Human Resources*, vol. 50, no. 4, pp. 873–909.

Gan, DL 2009. 'Energy service companies to improve energy efficiency in China: barriers and removal measures', *Procedia Earth and Planetary Science*, vol. 1, no. 1, pp. 1695–1704.

Gao, X & Zhu, ML 2010, '*Woguo zhengfu bumen jian xietiao wenti tantao (The investigation on the coordination between government departments in China)*', *Chengdu Xingzheng Xueyuan Xuebao*, vol. 1, pp. 4–7.

Gonzalez, G 2014, *China lets carbon programs blossom, nut will they take root?*, viewed 20, February 2017, <http://www.ecosystemmarketplace.com/articles/china-lets-carbon-programs-blossom-but-will-they-take-root/>.

Goron, C & Cassisa, C 2016, 'Regulatory institutions and market-based climate policy in China', *Global Environmental Politics*, vol. 17, no. 1, pp. 99–120.

Goron, C & Cassisa, C 2017, 'Regulatory Institutions and Market-Based Climate Policy in China', *Global Environmental Politics*, vol. 17, no. 1, pp. 99–120.

Hartmann, P & Apaolaza-Ibáñez, V 2012, 'Consumer attitude and purchase intention toward green energy brands: The roles of psychological benefits and environmental concern', *Journal of Business Research*, vol. 65, no. 9, pp. 1254–1263.

Huo, YA 2018, *Guowuyuan xinyilun jigou gaige zuixin jiedu: fagaiwei "shoushen" (State Council's new wave of institution reform: NDRC's "weight loss")*, viewed 21 October 2018, <http://news.sina.com.cn/c/nd/2018-03-14/doc-ifyscsmv3318743.shtml>.

Hvistendahl, M 2008, 'China's Three Gorges Dam: an environmental catastrophe?', *Scientific American*, viewed 20 December 2016, <https://www.scientificamerican.com/article/chinas-three-gorges-dam-disaster/>.

Jackson, S & Sleigh, A 2000, 'Resettlement for China's Three Gorges Dam: socio-economic impact and institutional tensions', *Communist and Post-Communist Studies*, vol. 33, no. 2, pp. 223–241.

Jiang, HQ & Li QZ 2013, 'Hetong nengyuan guanli fazhan qushi yu shangye yinhang duice jianyi (The development trend of EPC and recommendations for commercial banks' strategies)', viewed 7 July 2017, <http://finance.sina.com.cn/money/bank/20130712/095816103463.shtml>.

Kostka, G & Shin, K 2013, 'Energy conservation through energy service companies: empirical analysis from China', *Energy Policy*, vol. 52, pp. 748–759.

Kumar, D, Hossain, A & Gope, M.C 2015, 'Role of micro credit program in empowering rural women in Bangladesh: a study on Grameen Bank Bangladesh Limited', *Asian Business Review*, vol. 3, no. 4, pp. 114–120.

Lee, CK 2000, 'The revenge of history: collective memories and labor protests in North-Eastern China', *Ethnography*, vol. 1, no. 2, pp. 217–237.

Leung, JC 1995, 'The political economy of unemployment and unemployment insurance in the People's Republic of China', *International Social Work*, vol. 38, no. 2, pp. 139–149.

Li, J & Colombier, M 2009, 'Managing carbon emissions in China through building energy efficiency'. *Journal of Environmental Management*, vol. 90, no. 8, pp. 2436–2447.

Li, Y 2012, 'AHP-Fuzzy evaluation on financing bottleneck in energy performance contracting in China', *Energy Procedia*, no. 14, pp. 121–126.

Liu, R 2014, *Bejing shizhang li shengsizhuang: zhibuhao kongqi wuran "titou laijian" (Beijing Mayor took duty pledge: give up life if air pollution is not managed well)*, viewed 7 July 2017, <http://news.qq.com/a/20140118/007062.htm>.

Lo, AY & Cong, R 2017, 'After CDM: domestic carbon offsetting in China', *Journal of Cleaner Production*, vol. 141, pp. 1391–1399.

Lo, AY & Yu, X 2015, 'Climate for business: opportunities for financial institutions and sustainable development in the Chinese carbon market', *Sustainable Development*, vol. 23, no. 6, pp. 369–380.

Lu, XX 2015, '*Shanghai hetong nengyuan guanli jin qianjia, tansuo jieneng fuwu moshi (Shanghai EPC entering thousands of companies, exploring the EPC service model)*', viewed, 10 October 2017, <http://finance.chinanews.com/ny/2015/04-28/7239027.shtml>.

Ma, MP 2016, 'Woguo zaijiuye peixun de xianzhuang fenxi (Analysis of China's current reemployment situation)', *Renli Ziyuan Guanli*, vol. 5, pp. 249–249.

MEE 2018, *Yingdui qihoubianhua si (Department of Climate Change, MEE)*, viewed 27 October 2018, <http://www.mee.gov.cn/xxgk2018/xxgk/zjjg/jgsz/201810/t20181008_644817.html>.

MOHRSS 2018 *Shiye baoxianjin shenling fafang banfa (The application and allocation methods for unemployment payment insurance payment)*, viewed 29 December 2018, <http://www.mohrss.gov.cn/SYrlzyhshbzb/zcfg/flfg/gz/201901/t20190103_308094.html>.

Mueller, B, Berg, M, Yao, ZP, Zhang, XF, Wang, D & Pfluger, A 2008, 'How polluted is the Yangtze River? Water quality downstream from the Three Gorges Dam', *Science of the Total Environment*, vol. 402, no. 2, pp. 232–247.

New, T & Xie, Z 2008, 'Impacts of large dams on riparian vegetation: applying global experience to the case of China's Three Gorges Dam', *Biodiversity and Conservation*, vol. 17, no. 13, pp. 3149–3163.

Ng, J 2013, *Blocked on Weibo: what gets suppressed on China's version of Twitter (and why)*, The New Press, New York.

Nickell, S 1997, 'Unemployment and labor market rigidities: Europe versus North America', *The Journal of Economic Perspectives*, vol. 11, no. 3, pp. 55–74.

Pauliuk, S, Dhaniati, NMA & Müller, DB 2012, 'Reconciling sectoral abatement strategies with global climate targets: the case of the Chinese passenger vehicle fleet', viewed 11 March 2017, <https://brage.bibsys.no/xmlui/bitstream/handle/11250/2393648/Energy_Cars_China-Self-archive.pdf?sequence=1>.

Qianzhan Industrial Research Institution 2015, *China Energy Management Contract (EMC) industry development prospects and investment strategy report (2015-2020)*, viewed 5 May 2017, <http://bg.qianzhan.com/report/detail/b84b3f8f853d4362.html>.

Richards, D & Smith, M 2006, 'The tension of political control and administrative autonomy: from NPM to a reconstituted Westminster model', in T Christensen & P Lægreid (eds), *Autonomy and control: coping with agencies in the modern state*, Edward Elgar, Cheltenham, pp. 181–202.

Shankleman, J 2017, 'Xi at Davos urges Trump to stay in "Hard Won" Paris climate deal', *Bloomberg*, viewed 1 May 2017, <https://www.bloomberg.com/news/articles/2017-01-16/climate-experts-see-xi-touting-clean-energy-leadership-at-davos>.

Shen, LN 2015, *Yi zhang tu rangnin cedi kandong CCER shichang (One chart for you to understand CCER market)*, viewed 10 Feb 2017, <http://www.tanpaifang.com/CCER/201502/0442138.html>.

Shen, W 2015, 'Chinese business at the dawn of its domestic emissions trading scheme: incentives and barriers to participation in carbon trading', *Climate Policy*, vol. 15, no. 3, pp. 339–354.

Stanway, D 2017, 'UPDATE 1-China renewable power waste worsens in 2016 – Greenpeace', *Global Energy News*, viewed 1 May 2017, <http://uk.reuters.com/article/china-renewables-waste-idUKL3N1HR1HP>.

Stanway, D 2018, 'China regions accused of faking pollution compliance in new probe', *Reuters*, viewed 4 December 2018, <https://af.reuters.com/article/worldNews/idAFKCN1MW0PN>.

State Council of PRC 2016, *Guowuyuan guanyu yinfa "shisanwu" jieneng jianpai zonghe gongzuo fang'an de tongzhi guofa [2016] 74 hao (The State Council's notification on the comphrehensive work plan of ECERS in 13th Five-Year Plan [2016 State Council Office Document no. 74])*, viewed 14 May 2017, <http://www.gov.cn/zhengce/content/2017-01/05/content_5156789.htm>.

Tanjiaoyi 2018, *CCER xiangmu shifou jinru fuhuoqi? (Is CCER program revived?)*, viewed 29 November 2018, <www.tanjiaoyi.com/article-24172-1.html>.

Tatsiramos, K & Ours, JC 2014, 'Labor market effects of unemployment insurance design', *Journal of Economic Surveys*, vol. 28, no. 2, pp. 284–311.

Taylor, RP 2009, 'Achieving sustainability in World Bank energy efficiency projects: lessons learned developing ESCOs in China', *Strategic Planning for Energy and the Environment*, vol. 29, no. 1, pp. 32–41.

The Central People's Government of PRC 2005, *Shiye baoxian tiaoli, guowuyuan ling di 258 hao (Unemployment Insurance Regulation, The State Council No. 258 Announcement)*, viewed 2 February 2017, <http://www.gov.cn/banshi/2005-08/04/content_20258.htm>.

Toscano, N 2016, 'Ford shutdown: Auto manufacturing worker fear for their future', *Sydney Morning Herald*, viewed 1 October 2016, <http://www.smh.com.au/business/workplace-relations/ford-shutdown-auto-manufacturing-workers-fear-for-their-futures-20160929-grrxfy.html>.

Vodopivec, M & Tong, MH, 2008, *China: improving unemployment insurance*. Social Protection & Labor, World Bank, Washington, DC.

Wang, J, Cooke, FL & Lin, Z 2016, 'Informal employment in China: recent development and human resource implications', *Asia Pacific Journal of Human Resources*, vol. 54, no. 3, pp. 292–311.

Wang, J & Yang, F 2014, 'Qiantan huanbao yishi yu xiaofeizhe xingwei (Discussion on environmental conciousness and consumer behavior)', *Jieneng Huanbao*, viewed 11 July 2016, <http://www.doc88.com/p-9099610726176.html>.

Wang, S 2016, *Sheung yixia huanbao bumen jiang chuizhi guanli, 12 shengfen tichu shidian shenqing (Provincial and under environmental protections bureaus will implement direct supervision, 12 provinces applied to participate in the pilot program)*, viewed 12 February 2019, <http://www.xinhuanet.com/politics/2016-09/23/c_129295639.htm>.

Wang, XP 2017, *Jiuye dalibao zhuli shiye renyuan zaijiuye (Reemployment packages help the unemployed to gain reemployment)*, viewed 7 July 2017, <http://career.youth.cn/zxzx/201706/t20170622_10143442.htm>.

Wu, MY 2009, 'Xin xingshi xia zaijiuye peixun cunzai de wenti ji duice (The problems associated with re-employment under the new circumstances and solutions)', *Journal of Shengli College China University of Petroleum*, vol. 23, no. 1, pp. 58–60.

Xie, P 2003. 'Three-Gorges Dam: risk to ancient fish', *Science*, vol. 302, no. 5648, pp. 1149–1151.

Xinhua News 2014, *Taotai luohou channeng, gongren hequ hecong (Phasing out outdated production capacity, where should the workers go)*, viewed 11 April 2017, <http://news.xinhuanet.com/mrdx/2014-06/24/c_133433050.htm>.

Xu, BN & Albert, E 2017, 'Media Censorship in China', *Council on Foreign Relations*, viewed 20 November 2018, <https://www.cfr.org/backgrounder/media-censorship-china>.

Zhihu 2015, *Boda 12369 huanbao rexian youyongma? (Does it work to dial 12369 environmental protection hotline?)*, viewed 10 May 2017, <https://www.zhihu.com/question/28437244>.

Zhongchuang Tantou 2015, 'Qida tanjiaoyi shichang CCER zhongjie zhengce zongjie (A summary of CCER policies of China's seven carbon trading markets)', *CCER*, viewed 15 February 2016, <http://www.tanpaifang.com/CCER/201507/1645858.html>.

Zhou, K 2014, 'Xiaofeizhe gen yuanyi wei you huanbao he shehui zeren yishi de qiye maidan (Consumers are more willing to deal with companies with environmental consciousness and social responsibilities)', *China Youth News*, viewed 11 July 2016, <http://zqb.cyol.com/html/2014-06/23/nw.D110000zgqnb_20140623_3-10.htm>.

Zhu, Y 2018, *Quanguo tanshichang jiyi qidong, sheli tanshui hai you biyao ma? (National carbon market is already set up, is it still necessary to have carbon tax?)*, viewed 17 November 2018, <http://www.sohu.com/a/271712352_660963>.

7 Taking stock

Achievements and the road ahead

To answer the question of how China can optimise its current energy policies in order to achieve better outcomes in climate change mitigation, the book conducted analyses on three objects at two levels. At the policy development level, it investigated the role of and relationship between China's major actors in climate-related energy policy. At the policy implementation level, it evaluated the strengths and weaknesses of CDM covering the energy-related projects and Energy Conservation and Emission Reduction Scheme (ECERS) in the energy sector during the 11th and 12th Five-Year Plans. The strength in the government's administrative power was worthy of affirmation. It was especially effective in the closing down of small coal-fired power plants. The government's power was also reflected in the utilisation of resources for rapid, large-scale renewable energy development. However, this book also revealed a range of obstacles in China's institutional arrangements and policy implementation process. It concluded that China's climate-related energy policies are bold and ambitious. The major problem associated with the policies lay neither in speed nor in scale, but in coordination, efficiency, integrity, and quality.

During the 13th Five-Year Plan, the government actor showed the most significant updates. Ministry of Environmental Protection (MEP) was renamed Ministry of Ecology and Environment (MEE), and Department of Climate Change (DCC) was officially moved under MEE in 2018. China's climate change affair was eventually managed by an environmental ministry. Nonetheless, fragmentation and overlapping policy responsibilities still exist among relevant climate-related ministries and departments. Adopting a whole-of-government (WG) approach is recommended to eliminate conflicts, enhance coordination, and conduct more effective climate and energy policies. The relationship between the central, provincial, and municipal governments in climate change mitigation and general environmental protection is even more problematic. It is difficult for the central government to monitor policy implementation at the out-of-reach municipal level, where the municipal government usually puts economic development as its top priority for tax revenues. In the 13th Five-Year Plan, MEE initiated pilot program for provincial environmental protection bureaus (EPBs) to carry out direct supervision for EPBs at lower administrative levels. This was a very positive move because provincial and municipal EPBs' actions against pollution are no longer restricted by provincial and municipal governments. If the pilot program proved successful, this initiative is expected to be carried out nationwide.

Enterprises including state-owned enterprises (SOEs) and non-SOEs were also forces that shaped China's climate-related energy policies. Large energy SOEs generally had more bargaining power with the central government; while smaller non-SOEs were weaker. The expert community contributed greatly to the input of climate-related policy development. This was mainly due to the lack of expertise in the climate change domain within the government. The media faced great political obstacles when reporting serious environmental problems. This status worsened under the Xi's government. Environmental Non-governmental Organisations (ENGOs) and the general public played a limited role in the formation of China's climate-related policies. Their contribution was mainly through the channels of media and public online forums, which could form only a tangential policy input. This investigation showed that the central government indeed played the most important role in China's climate-related energy policies. Therefore, how well the government designed the policy had the most significant impact on policy outcomes.

Nonetheless, policy design was only one aspect that influences the outcome of the policy. The quality of the implementation had even bigger influence. ECERS shut down over 100 million kW of small coal-fired power capacity during the 11th and 12th Five-Year Plans. This outcome was impressive; however, sub-optimal elements were also numerous. Unfair and delayed compensations to some of the laid-off workers and private small coal power plant owners posed great difficulty for those affected. The corruption and fraudulent claims for closure compensation seriously damaged the policy's integrity, especially in ascertaining the real level of emissions reduction. However, replacing the small coal power plants with more efficient large ones improved overall efficiency in the coal power sector. The policy fulfilled the energy conservation function, but it had little impact on overall emissions reduction in the coal power sector. Also, it did not fundamentally optimise China's energy structure. This would require the government to reconsider its current planning of coal-fired power plants and make some serious decisions in cutting down coal usage and replacing it with renewable energies.

A large proportion of ECERS' investment went to the development of the renewable energy industry. China's total renewable energy capacity reached a quarter of the world's total and became the world's leader in renewable energy capacity. However, serious challenges were evident in the renewable energy sector. Within the renewable energy industry, unfair competition between large SOEs and smaller private companies seriously restricted the enthusiasm of private investment in the sector. Large SOEs only passively included renewable energy projects in their portfolio because the government required them to do so. Although they received many types of government financial assistance for their renewable energy projects, their primary interest was still in the traditional fossil-based energies. They basically use their renewable energy projects as bargaining power for future government approval of fossil-based power plants. To the contrary, smaller private companies received little support in comparison. This unfair treatment discouraged private investment in the renewable energy industry and was not beneficial for the sustainability of long-term renewable energy development.

An even more serious problem in China's renewable energy industry was the waste of wind and solar power. In many northern provinces, wind and solar power plants were not allowed to operate at full capacity under the provincial and municipal governments' administrative pressure. Excuses such as limited grid capacity and limited consumption capacity contributed to the wastage of up to 70% of wind power in some regions (BJX Wind Power 2016, n.p.). This situation was caused by the competition of interests from the coal mining industry, coal-fired power plants, grid companies, and provincial governments. China's wind-rich regions largely overlapped with China's coal-rich regions. The coal mining industry lobbied the provincial governments to approve more coal-fired power plants to sustain their long-term operation. Some provincial governments did so to gain immediate and sustained tax revenue. Meanwhile, for the same purpose of raising tax revenue, they also approved more wind and solar power capacity for the purpose of keeping wind turbine and solar panel manufacturing industry in their provinces. The blind expansion of both coal and renewable energy capacity creates energy surplus in the less developed northern and western regions. Nonetheless, if the renewable energy law was followed and priority was given to renewable energy, the vast waste of wind and solar energy in northern China would have not happened. Thus, the book concluded that the problem associated with China's climate-related energy policies was neither a scale problem nor a speed problem, but a problem of integrity, efficiency, and quality.

In order to overcome the weaknesses in China's coal power and renewable power sectors and improve the integrity, efficiency, and quality of ECERS, this book identified several implications. It included adjusting Unemployment Insurance (UI) duration and providing small business loans and retraining scholarships for unemployed workers of the shut-down small coal-fired power plants; abolishing cash rewards for energy conservation and emission reduction and replacing it with tax measures; toughening the penalty for violating the *Renewable Energy Law*; appointing the head of provincial and municipal governments to be responsible for the renewable energy sector within their authority, and any amount of wasted renewable energy will be deducted from their tax income in cash value; utilising renewable energy in electric car charging facilities; and prioritising wind power as a source for winter central heating in northern provinces. It is hoped these recommendations will be able to address the weaknesses of current policy and point out new directions for future policy development.

Energy Performance Contracting (EPC) was another component that this book chose to investigate under ECERS. EPC was a market mechanism which targeted energy conservation and efficiency improvement by utilising better technologies or management procedures or both. EPC was relatively new to China. As a new phenomenon, it encountered several obvious challenges, including the lack of trust among EPC stakeholders, lack of involvement of formal finance, and difficulty in finding the compatible clients or energy service companies (ESCOs). In response to these issues, this book suggested introducing a unified national accreditation system for ESCOs; conducting special EPC industry report targeting potential financial providers; and creating web-based info-exchange platforms for ESCOs and clients. To further promote the development of EPC, the book also suggested qualified ESCOs to undertake energy efficiency CCER projects. It was demonstrated that the expertise and experience of such ESCOs would have much to offer in China's national carbon market.

Clean Development Mechanism (CDM), another market mechanism analysed in this book, introduced great climate change mitigation concept, procedures, and system to China. It brought China a range of benefits in sustainable development. It could also be viewed as an inspiration for China's domestic carbon market. ECERS was adopted in a mainly top-down administrative fashion. Compared with CDM, it was expensive and less cost efficient. China's national ETS had been implemented for a year at the time of this book's completion. Whether or not it will dramatically improve the cost-effectiveness of climate change mitigation still requires time to evaluate. Currently, only the electricity power sector was included in the national ETS. As more industries enter the scheme, the operation standardises, and the market matures, it will provide an opportunity for the comprehensive analysis of the overall efficiency and effectiveness of China's national ETS. It is certain that the Chinese government started to realise the importance of addressing climate change and atmospheric pollutions. Therefore, better energy policies with a focus on climate stability can be expected.

Final remarks

It has been over a decade since China implemented its first ECERS. Much has happened in China's climate-related policy domain. China participated in the international carbon offsetting program CDM, completed its carbon trading pilot program, and implemented the national ETS at the end of 2017. The ECERS of the 13th Five-Year Plan (2016–2020) was expected to be promising as it included a clear message on dramatically reducing coal consumption and addressing the wasting of wind and solar power; however, it has not seen dramatic improvement up till this stage.

Many of the implications drawn out of the analyses of this book provide innovative and workable solutions to the current policies. Directions on how China's emission mitigation can be improved through optimising current energy policies were demonstrated. More importantly, the data used in this book was not limited to official sources, but it also incorporated the views of Chinese officials, scholars, energy company managers, ENGO personnel, and the general public. Therefore, it has provided a fuller picture of the analysed policies than currently exist. As a result, the findings in this book are valuable for China's current and future climate-related energy policies.

Index